MATH FOR WATER TREATMENT OPERATORS

MATH FOR WATER TREATMENT OPERATORS

Practice Problems to Prepare for Water Treatment Operator Certification Exams

John Giorgi

FIRST EDITION

American Water Works Association

Disclaimer

Although this study guide has been extensively reviewed for accuracy, there may be an occasion to dispute an answer, either factually or in the interpretation of the question. Both AWWA and the author have made every effort to correct or eliminate any questions that may be confusing or ambiguous. If you do find a question that you feel is confusing or incorrect, please contact the AWWA Publishing Group.

Additionally, it is important to understand the purpose of this study guide. It does not guarantee certification. It is intended to provide the operator with an understanding of the types of math questions he or she will be presented with on a certification exam and the areas of knowledge that will be covered. AWWA highly recommends that you make use of the additional resources listed at the end of this study guide in preparing for your exam.

Printed in the United States of America
ISBN-10 1-58321-454-2
ISBN-13 978-1-58321-454-1

Library of Congress Cataloging-in-Publication Data

Giorgi, John.

Math for water treatment operators / John Giorgi and prepared by the editors of American Water Works Association.

p. cm.

Includes bibliographical references and index.

ISBN 1583214542

1. Water--Purification--Mathematics. 2. Water--Purification--Problems, exercises, etc. 3. Engineering mathematics--Formulae. I. American Water Works Association. II. Title.

TD433.G56 2006
628.101'51--dc22

2006047723

American Water Works Association

6666 West Quincy Avenue
Denver, Colorado 80235-3098
303.794.7711

This work is dedicated to my wife, Flora Zhou Giorgi, my children, Sara, Stephanie, and Steve; my mother, Thelma Giorgi, and my father, Albert Peter Giorgi.

CONTENTS

TABLES

ACKNOWLEDGMENTS

I would like to thank the staff and editors of the American Water Works Association, Mary Kay Kozyra, Colin Murcray, John Stubbart, and Bill Lauer; and the reviewers of the manuscript: Kevin Linder, Wemlinger Water Treatment Plant, Aurora, Colo.; Sid Copeland, City of Louisville Water Plant, Louisville, Colo.; and Ron Mathis, City of Golden, Colo.; for their help and guidance in making this book possible. Their assistance is greatly appreciated.

I would also like to thank fellow water treatment operators at El Dorado Irrigation District in California: Kurt Mikkola, William Petterson, Radenko Odzakovic, Robin Peck, Dave Herrmann, Aaron Davis, Zol Whitman, Jon Beale, Mike Royere, Jake Maker, Dean Bugher, Michael Ragan, Rick Talone, and Boyd Haycock.

I am grateful to my wife Flora and my children, Steve, Stephanie, and Sara. Their patience and support for my long hours working on this book is greatly appreciated.

PREFACE

The first edition of *Math for Water Treatment Operators* was written to provide students and operators with examples of a variety of different problems that will be encountered both on certification exams and on the job. This book is divided into four parts consisting of material associated with the first four grade levels for treatment. The problems increase in difficulty as the grade number increases.

Each problem is presented with easily followed steps and comments to facilitate understanding. One possible way to go through the math problems presented in this study guide is for you to cover the page you are working on with a piece of paper or cardboard. Then slowly move the cover down until you can read the question. Do the problem on a separate piece of paper. Uncover the worked solution and compare your method and result to the book's method and result. If your answer is the same, but your method is different, that's OK. Remember that there may be more than one way to solve a problem. If there is a certain problem that gives you trouble, try to do the problem again on another day until you completely understand it. Do similar problems that may be found in the other grades in the book. The more math problems you do, the more comfortable you will become with them.

Included after each grade level are tests to help you determine where your strengths and weaknesses are. Each test consists of randomly chosen problems from the associated problems in that chapter. The test problems are followed immediately by the solution. You can complete the problem on a separate piece of paper. After completing the test, the procedures and answers can be checked against the provided procedures and solutions.

Comprehensive CT tables are provided on a CD-ROM disk in Microsoft Excel format. Common conversion factors are included in Appendix A for reference purposes and for doing problems in this book. Appendix B is a summary of the formulas used in this book. Appendix C consists of chemistry tables that are needed for some of the problems. Appendix D includes US Environmental Protection Agency maximum contaminant level (MCL) tables for reference. And Appendix E lists the abbreviations used in this book.

Any suggestions for improving this math book including additional types of problems would be appreciated by the author. Please send your suggestions or questions to John Giorgi in care of AWWA, 6666 Quincy Avenue, Denver, CO 80235-3098.

INTRODUCTION

SIGNIFICANT FIGURES

When you see an answer to a mathematical problem, laboratory test result, or other measured value, do you ever wonder how accurate it is? An answer's accuracy is based on the accuracy of the values used to determine the answer; the accuracy of a laboratory result depends on the precision of the measuring instruments, and even the laboratory analyst.

The following discussion explains how to determine the number of significant figures or digits an answer to any particular problem should have, i.e., how many decimal places, if any, should the answer have.

The number 30.03 has four significant figures, and the number 33,000 has only two. Why is this so? The number 30.03 has been measured to the hundredth place so the zeroes that are straddled by the threes are significant. In fact, all figures to the left of a decimal point are significant (e.g., 7.000 has four significant figures). The second number, 33,000 has only two significant figures because the zeroes are only placeholders and are thus not significant. (See the following exercise on significant figures.)

"Rounding Off"

"Rounding off" numbers is simply the dropping of figures starting on the right until the appropriate number of significant figures remains. Let's look at the three rules and an example for each that governs the process of rounding numbers.

1. When a figure less than 5 is dropped, the next figure to the left remains unchanged. Thus, the number 11.24 becomes 11.2 when the 4 is dropped.
2. When the figure is greater than 5, that number is dropped and the number to the left is increased by 1. Thus, 11.26 will become 11.3.
3. When the figure that needs to be dropped is a 5, round to the nearest even number. This prevents rounding bias. Thus, 11.35 becomes 11.4 and 46.25 becomes 46.2.

The Significance or Insignificance of Zero

A zero may be a significant figure if it is a measured value, or a zero may be insignificant and serve only as a spacer for locating the decimal point. If a zero or zeroes are used to give position value to the significant figures in the number, the zero or zeroes are not significant. An example of this would be the following expression: 1.23 mm = 0.123 cm = 0.000123 m = 0.00000123 km. The zeroes are insignificant and only give the significant figures, 1 2 3, a position that dictates their value.

Addition and Subtraction

When adding and subtracting, only similar units that are written to the same number of decimal places may be added or subtracted. Also, the number with the fewest number of decimal places, and not necessarily the fewest number of significant figures, places a limit on the number that the sum can justifiably carry.

For example, add the following numbers: 446 mm + 185.22 cm + 18.9 m. First, convert the quantities to similar units, which in this case is the meter (second row below). Next, choose the least accurate number, which is 18.9. It has only one number to the right of the decimal so the other two values will have to be rounded off (third row below).

446 mm	=	0.446 m	=	0.4 m
185.22 cm	=	1.8522 m	=	1.8 m
18.9 m	=	18.9 m	=	18.9 m
				21.1 m

When adding numbers (including negative numbers), the rule is that the least accurate number will determine the number reported as the sum. In other words, the number of significant figures reported in the sum cannot be greater than the least significant figure in the group being added. Another example is given below in which the least accurate number, 170, dictates how the other three numbers will have to be changed before addition is done.

1.023 g	=	1 g
23.22 g	=	23 g
170 g	=	170 g
1.008 g	=	1 g
195.251 g		195 g

However, you cannot report either of these values. The third value, 170 g, has two significant numbers, while all the others have four. The limiting factor is this third value, 170 g. The number 195 has three significant figures and, thus, cannot be used either. The answer must be reported as 200 g even though this looks wrong because it only has one significant figure!

Multiplication and Division

When multiplying or dividing, the number that has the fewest significant figures will dictate how the answer will be written. Suppose we have to multiply two numbers: (23.88)(7.2) = 171.936. The first number has four significant figures, while the second has only two. The answer should only be written with two significant figures, as 170, because one of the numbers, 7.2, has only two significant figures. In both multiplication and division, "rounding off" never should be done before the mathematical exercise. Only the result should be appropriately "rounded off."

In the following exercise, give the number of significant figures from each of the values below:

	VALUE	ANSWER
a.	8.34 lb/gal	3
b.	0.04 ntu	1
c.	23.06 mgd	4
d.	5 1-ton chlorine containers	infinite
e.	5.50 mg/L	3
f.	0.00000254	3
g.	50,000 tons	1
h.	6.023×10^{23} (the number of atoms in a mol, Avogadro's number)	4
i.	43,560 ft^3/acre-ft	infinite
j.	340 m	2
k.	5.01 g	3
l.	1,200,050 gal	6
m.	1,000,000/M	infinite
n.	1,440 min/d	infinite
o.	7.481 gal/ft^3	4
p.	86,400 s/d	infinite

ROUNDING IN THIS BOOK

The problems in this book are done in steps so you can see each operation visually, which facilitates understanding. In so doing, a certain amount of "pre-rounding" has to occur, otherwise the numbers continue, in most cases, to absurdity. This "pre-rounding" was done, as much as possible, so that the final answer would not be affected.

SET-UP OF PROBLEMS IN THIS BOOK

Note that many of the same types of problems have the equations written each time and that the units are written throughout each problem. This is done to help you form a good habit of writing down the equations and units. This habit should be carried over to taking certification tests. In some cases, partial credit will be given for a correctly written equation. When the units are written down, it will not only help you set up the problem correctly by seeing how the units will cancel, but will also make it easier for the person correcting the test to see your intent and also correct the problems.

Pi (π)

The number used for pi (π) will be 3.14 throughout this book.

DERIVATION OF THE NUMBER 0.785

The number 0.785 is used extensively in this book in conjunction with the diameter squared. Examples include determination of the area of a circle or the volume of a tank. The same answer can be achieved using πr^2, where "r" equals the radius. But how was this number, 0.785, derived?

The following applies:

1. **$(0.785)(\text{Diameter})^2 = \pi r^2$**
2. **$\text{Diameter}^2 = 4r^2$**

Assume x is this number (0.785) and is not yet known. We know some number, x, times the diameter squared equals π times the radius squared. The equation is:

$$x(\text{Diameter})^2 = \pi r^2$$

From 1 above and substituting x for (0.785),

substitute $4r^2$ for Diameter^2 from 2 above:

$$x(4r^2) = \pi r^2$$

Solve for x,

$$x = \frac{\pi r^2}{4r^2} = \frac{\pi}{4}$$

Substitute 3.14 for π and solve:

$$x = \frac{3.14}{4} = \mathbf{0.785}$$

Thus, x is equal to 0.785, which is what we wanted to prove.

Most water treatment operators like to use $(0.785)(\text{Diameter})^2$, while engineers and scientists like to use πr^2. Because both will be encountered, it is advisable to know both methods.

CHAPTER 1

WATER TREATMENT

Grade 1

Students preparing for grade 2 through grade 4 water treatment certification tests should also understand these problems.

CONVERSION PROBLEMS

1. How many pounds are there in 1 ft^3 of water?

(8.34 lb/gal)(7.48 gal/ft^3) = 62.3832 lb/ft^3, round to **62.4 lb/ft^3**

2. How much does exactly 100 gal of water weigh?

The equation is:

$$\textbf{(Number of gal)}\ \frac{\textbf{(8.34 lb)}}{\textbf{gal}} = \textbf{number of pounds}$$

$$(100\ \text{gal})\ \frac{(8.34\ \text{lb})}{\text{gal}} = \textbf{834 lb}$$

Note: In this case 100 gal is not a measurement, but an exact amount. It therefore has an infinite number of significant figures. The limiting factor in this problem is 8.34 lb/gal (three significant figures).

3. Convert 8.2 ft³/s to gallons per minute.

$$(8.2\ \text{ft}^3\text{/s})(60\ \text{s/min})(7.48\ \text{gal/ft}^3) = 3{,}680\ \text{gpm, round to } \mathbf{3{,}700\ gpm}$$

4. How many gallons are there in 82 ft³?

$$\frac{(7.48\ \text{gal})}{1\ \text{ft}^3}(82\ \text{ft}^3) = 613.36\ \text{gal, round to } \mathbf{610\ gal}$$

5. Convert 2,445 gal to cubic feet.

$$(2{,}445\ \text{gal})\frac{(1\ \text{ft}^3)}{7.48\ \text{gal}} = 326.87\ \text{ft}^3\text{, round to } \mathbf{327\ ft^3}$$

Note**:** The limiting significant figure is 7.48 gal.

6. How much does 725 gal of water weigh in pounds?

$$(725\ \text{gal})\frac{(8.34\ \text{lb})}{\text{gal}} = 6{,}046.5\ \text{lb, round to } \mathbf{6{,}050\ lb}$$

7. Convert 5.1 mgd to cubic feet per second.

$$(5.1\ \text{mgd})\frac{(1{,}000{,}000\ \text{gal})}{1\ \text{mil gal}}\frac{(1\ \text{ft}^3)}{7.48\ \text{gal}}\frac{(1\ \text{d})}{1{,}440\ \text{min}}\frac{(1\ \text{min})}{60\ \text{s}} = \mathbf{7.9\ ft^3/s}$$

8. Convert 15.0 acre-ft to cubic feet.

The equation is:

(Number of acre-ft) (43,560 ft³/acre-ft)

$$(15.0 \text{ acre-ft})(43{,}560 \text{ ft}^3/\text{acre-ft}) = 653{,}400 \text{ ft}^3, \text{ round to } \mathbf{653{,}000 \text{ ft}^3}$$

9. Convert 4,078,611 ft³ to acre-feet.

$$\frac{4{,}078{,}611 \text{ ft}^3}{43{,}560 \text{ ft}^3/\text{acre-ft}} = \mathbf{93.63202 \text{ acre-ft}}$$

10. Convert 11.9 mgd to cubic feet per second.

$$(11.9 \text{ mgd}) \frac{(1{,}000{,}000 \text{ gal})}{1 \text{ mil gal}} \frac{(1 \text{ ft}^3)}{7.48 \text{ gal}} \frac{(1 \text{ d})}{1{,}440 \text{ min}} \frac{(1 \text{ min})}{60 \text{ s}} = \mathbf{18.4 \text{ ft}^3/\text{s}}$$

11. Convert 5.6 ft³/s to gallons per minute.

$$(5.6 \text{ ft}^3/\text{s})(60 \text{ s/min})(7.48 \text{ gal/ft}^3) = 2{,}513.3 \text{ gpm, round to } \mathbf{2{,}500 \text{ gpm}}$$

12. Convert 3.2 ft³/s to million gallons per day.

$$\frac{(3.2\ \text{ft}^3)}{\text{s}}\ \frac{(60\ \text{s})}{\text{min}}\ \frac{(1{,}440\ \text{min})}{\text{day}}\ \frac{(7.48\ \text{gal})}{\text{ft}^3}\ \frac{(1\ \text{mil gal})}{1{,}000{,}000\ \text{gal}} = \mathbf{2.1\ mgd}$$

13. How many million gallons are there in 22 acre-ft?

You can do the problem in one step as follows:

$$(22\ \text{acre-ft})(43{,}560\ \text{ft}^3/\text{acre-ft})(7.48\ \text{gal/ft}^3)(1\ \text{M}/1{,}000{,}000) = \mathbf{7.2\ mil\ gal}$$

14. How many million gallons are there in 43,000 acre-ft?

$$(43{,}000\ \text{acre-ft})(43{,}560\ \text{ft}^3/\text{acre-ft})(7.48\ \text{gal/ft}^3)(1\ \text{M}/1{,}000{,}000) = \mathbf{14{,}000\ mil\ gal}$$

Write down all the units so you can see which units cancel each other out.

TEMPERATURE CONVERSION PROBLEMS

15. **Convert 45°F to degrees Celsius.**

The equation for Celsius is:

$$°C = \frac{5}{9}(°F - 32)$$

$$°C = \frac{5}{9}(45 - 32) = \frac{5}{9}(13) = \mathbf{7.2°C}$$

16. **Convert 350°C to degrees Fahrenheit.**

The equation for Fahrenheit is:

$$°F = (\frac{9}{5})°C + 32 \text{ or use } F = 1.8°C + 32°F$$

$$°F = (\frac{9}{5})350 + 32 = 662°F\text{, round to } \mathbf{660°F}$$

17. **Convert 75°F to degrees Celsius.**

The equation for Celsius is:

$$°C = \frac{5}{9}(75 - 32)\ \frac{5}{9}(43) = \mathbf{24°C}$$

PERCENT AND PARTS PER MILLION CONVERSION PROBLEMS

18. Convert a solution that has 80,147 ppm to percent.

A 1% solution = 10,000 ppm or 10,000 mg/L

$$\frac{80{,}147 \text{ ppm}}{10{,}000 \text{ ppm/1\%}} = \textbf{8.0147\% solution}$$

19. An alternative way to convert 80,147 ppm to percent is:

$$\frac{80{,}147 \text{ ppm} \times 100\%}{1{,}000{,}000/1\text{M}} = \textbf{8.0147\% solution}$$

20. A solution was found to be 1.3% alum. How many milligrams per liter of alum are in the solution?

By setting up a ratio, and knowing that a 1% solution has 10,000 mg/L, a 1.3% solution will have:

$$\frac{10{,}000 \text{ mg/L}}{1\%} = \frac{x}{1.3\%}$$

$$\frac{(1.3\%)(10{,}000 \text{ mg/L})}{1\%} = \textbf{13,000 mg/L alum}$$

PERCENT CALCULATIONS

21. If 238 is 58%, what is 100%?

Writing a ratio can solve this problem.

$$238/58\% = x/100\% \qquad x = \frac{(100\%)(238)}{58\%} = 410.3, \text{ round to } \mathbf{410}$$

An easier way to solve the problem is to know that the number for 100% must be larger than 238. If we divide by the decimal for 58%, we get the same answer, and it is a little faster to solve.

$$x = 238/0.58 = \mathbf{410}$$

22. If 154 is 100%, what is 72%?

We multiply 154 by 72%/100% because we know the number has to be smaller than 154.

$$x = (154)(72\%/100\%) = 110.9, \text{ round to } \mathbf{111}$$

23. What is the percent removal across a settling basin if the influent is 24.3 ntu and the effluent is 1.45 ntu?

The equation is:

$$\textbf{Percent ntu removal} = \frac{\textbf{Influent ntu} - \textbf{Effluent ntu}}{\textbf{Influent ntu}} \times \mathbf{100\%} \textbf{ or } \frac{\textbf{In} - \textbf{Out}}{\textbf{In}} \times \mathbf{100\%}$$

$$\text{Percent ntu removal} = \frac{(24.3 \text{ ntu} - 1.45 \text{ ntu})}{24.3 \text{ ntu}} \times 100\% = \frac{22.85 \text{ ntu}}{24.3 \text{ ntu}} \times 100\%$$

Remember: Always perform addition or subtraction within parenthesis before multiplication or division.

$$\text{Percent ntu removal} = 0.940 \times 100 = \mathbf{94.0\%\ removal\ of\ ntu}$$

24. If 0.5 lb of soda ash are mixed into 1 gal of water, what is the percent of soda ash in the slurry?

The following applies: 1 gal of water = 8.34 lb

$$\text{Percent soda ash} = \frac{(0.5\text{ lb})(100\%)}{0.5\text{ lb} + 8.34\text{ lb}} = \frac{(0.5\text{ lb})(100\%)}{8.84\text{ lb}}$$

= 5.7% slurry, round to **6% soda ash slurry**

AVERAGE CALCULATIONS

25. What is the average in pounds per day for chlorine used given the following data?

Mon.	Tues.	Wed.	Thurs.	Fri.	Sat.	Sun.
74	78	81	84	77	73	70

The equation is:

$$\textbf{Average } \mathbf{Cl_2} \textbf{ used, lb/d} = \frac{\textbf{Sum of } \mathbf{Cl_2} \textbf{ used each day, lb}}{\textbf{Total time, d}}$$

$$\text{Average } Cl_2 \text{ used, lb/d} = \frac{74 + 78 + 81 + 84 + 77 + 73 + 70}{7\text{ d}}$$

= 76.7 lb/d Cl_2 round to **77 lb/d Cl_2**

26. What is the average number of pounds of lime used per day given the following data? All measured values were to the nearest pound.

Mon.	Tues.	Wed.	Thurs.	Fri.	Sat.	Sun.
458	523	491	503	478	471	460

The equation is:

$$\textbf{Average lime used, lb/d} = \frac{\textbf{Sum of lime used each day, lb}}{\textbf{Total time, d}}$$

$$\text{Average lime used, lb/d} = \frac{458 + 523 + 491 + 503 + 478 + 471 + 460}{7\text{ d}}$$

$$= \textbf{483 lb/d lime}$$

AREA PROBLEMS

27. What is the area of a basin that is 225 ft long and 42 ft wide?

The equation for a rectangular area is:

$$\textbf{Area} = \textbf{(Length)(Width)}$$

$$\text{Area} = (225\text{ ft})(42\text{ ft}) = 9{,}450\text{ ft}^2\text{, round to }\mathbf{9{,}500\ ft^2}$$

28. What is the area of a circular pond if it has a diameter of 224 ft?

The equation is:

$$\textbf{(Area)}(\pi r^2)\textbf{, where } \pi = 3.14$$

First, find the radius.

$$\text{Radius} = \text{Diameter}/2 = 224/2 = 112$$

$$\text{Area of tank} = (3.14)(112 \text{ ft})(112 \text{ ft}) = 39{,}388.16 \text{ ft}^2\text{, round to } \mathbf{39{,}400 \text{ ft}^2}$$

29. What is the diameter of a tank if the surface area is 5,200 ft²?

The equation is:

$$\textbf{Area} = \mathbf{(0.785)(Diameter)^2}$$

Solve for the diameter by rearranging the equation as shown.

$$\text{Diameter}^2 = \frac{\text{Area}}{0.785} = \frac{5{,}200 \text{ ft}^2}{0.785} = 6{,}624 \text{ ft}^2$$

Then, take the square root of 6,624 ft^2.

$$\text{Diameter} = 81.38 \text{ ft, round to } \textbf{81 ft in diameter}$$

30. What is the area of a circular reservoir if it has a diameter of 680 ft?

The equation is:

$$\textbf{Area} = \mathbf{(0.785)(Diameter)^2}$$

$$\text{Area of tank} = (0.785)(680 \text{ ft})(680 \text{ ft}) = 362{,}984 \text{ ft}^2\text{, round to } \mathbf{360{,}000 \text{ ft}^2}$$

31. If the surface area of a rectangular drying bed is 18,125 ft^2 and one side measures 60.0 ft, what is the length of the other side?

The equation is:

$$\textbf{Area} = \textbf{(Length)(Width)}$$

$$\text{Length, ft} = \frac{\text{Area of } 18{,}125 \text{ ft}^2}{(60.0 \text{ ft})} = 302.08 \text{ ft, round to } \mathbf{302\ ft}$$

CIRCUMFERENCE PROBLEMS

32. What is the circumference of a tank that is 100 ft in diameter as measured to the nearest foot?

The equation is:

$$\textbf{Circumference} = \boldsymbol{\pi}\textbf{(Diameter)}$$

$$\text{Circumference} = (3.14)(100 \text{ ft}) = \mathbf{314\ ft}$$

33. What is the circumference of a clarifier if the radius is 28 ft?

$$\textbf{Circumference} = \mathbf{2\pi}\textbf{(radius) or } \mathbf{2\pi r}$$

$$\text{Circumference} = 2(3.14)(28 \text{ ft}) = 175.84 \text{ ft, round to } \mathbf{180\ ft}$$

VOLUME PROBLEMS

34. What is the number of cubic feet in a sedimentation tank that is 340 ft long, 28 ft wide, and 7.5 ft deep?

The volume equation for a rectangular basin is:

Volume = (Length)(Width)(Depth)

Volume = (L)(W)(D) = (340 ft)(28 ft)(7.5 ft) = 71,400 ft^3, round to **71,000 ft^3**

35. A circular tank has a radius of 16 ft and is 23.5 ft high at the spill point. What is the capacity of the tank in cubic feet?

The volume formula for a circular tank is:

Volume = πr^2(Height), where *r* is the radius.

Volume of tank in cubic feet = 3.14(16 ft)(16 ft)(23.5 ft) = 18,890 ft^3, round to **19,000 ft^3**

36. Calculate the volume in cubic feet for a 14.0-in. diameter pipeline that is 1,228 ft long.

First, convert the diameter to feet.

$$(14.0 \text{ in.}) \frac{(1 \text{ ft})}{12 \text{ in.}} = 1.167 \text{ ft (diameter)}$$

Then, convert the diameter to the radius.

Radius = Diameter/2 = 1.167 ft/2 = 0.5835 ft (radius)

The formula for the volume of a pipe in cubic feet is:

πr^2(Length) or **(0.785)(Diameter)^2(Length)**

Using the first equation, the volume = (3.14)(0.5835 ft)(0.5835 ft)(1,228 ft)

Volume, ft^3 = 1,312.8 ft^3, round to **1,310 ft^3**

Using the second equation, the volume = (0.785)(1.167 ft)(1.167 ft)(1,228 ft)

Volume, ft^3 = 1,312.8 ft^3, round to **1,310 ft^3**

37. How many gallons would be in the pipe for the problem above?

(1,310 ft^3)(7.48 gal/ ft^3) = 9,799 gal, round to **9,800 gal**

38. What is the volume of a conical tank, in cubic feet, that has a radius of 10.0 ft and a height of 17 ft?

The equation is:

Volume, ft^3 = $1/3\pi r^2$(Height or Depth)

Volume, ft^3 = 1/3(3.14)(10.0 ft)(10.0 ft)(17 ft) = 1,779 ft^3, round to **1,800 ft^3**

39. Calculate the volume in cubic feet for a 16-in. diameter pipeline that is 1,850 ft long.

First, convert the diameter to feet.

$$(16 \text{ in.}) \frac{(1 \text{ ft})}{12 \text{ in.}} = 1.33 \text{ ft (Diameter)}$$

The equation is:

$$\textbf{Volume, ft}^3 = \textbf{(0.785)(Diameter)}^2\textbf{(Length)}$$

$$\text{Volume, ft}^3 = (0.785)(1.33 \text{ ft})(1.33 \text{ ft})(1{,}850 \text{ ft})$$

$$\text{Volume, ft}^3 = 2{,}569 \text{ ft}^3\text{, round to } \mathbf{2{,}600 \text{ ft}^3}$$

40. Determine the volume in cubic feet for a pipe with a diameter of 1.5 ft and a length of 1.75 mi.

First, determine the number of feet in 1.75 mi.

$$(5{,}280 \text{ ft/mi})(1.75 \text{ mi}) = 9{,}240 \text{ ft}$$

$$\text{Volume, ft}^3 = (0.785)(1.5 \text{ ft})(1.5 \text{ ft})(9{,}240 \text{ ft}) = 16{,}320.15 \text{ ft}^3\text{, round to } \mathbf{16{,}000 \text{ ft}^3}$$

41. A tank is conical at the bottom and cylindrical at the top. If the diameter of the cylinder is 18 ft with a depth of 32 ft and the cone depth is 12 ft, what is the approximate volume of the tank in cubic feet and gallons?

First, find the volume of the cone in ft^3.

$$\textbf{Volume, ft}^3 = \mathbf{1/3\pi r^2}\textbf{(Depth)}$$

where the radius = Diameter/2 = 18 ft/2 = 9.0 ft

$$\text{Volume, ft}^3 = 1/3(3.14)(9.0 \text{ ft})(9.0 \text{ ft})(12 \text{ ft}) = 1{,}017.36 \text{ ft}^3$$

Next, find the volume of the cylindrical part of the tank.

$$\text{Volume, ft}^3 = \pi r^2(\text{Depth}) = (3.14)(9.0 \text{ ft})(9.0 \text{ ft})(32 \text{ ft}) = 8{,}138.88 \text{ ft}^3$$

Then, add the two volumes for the answer.

$$\text{Total volume, ft}^3 = 1{,}017.36\ \text{ft}^3 = 8{,}138.88\ \text{ft}^3 = 9{,}156.24\ \text{ft}^3\text{, round to } \mathbf{9{,}200\ ft^3}$$

To find the number of gallons, multiply the total number of cubic feet by 7.48 gal/ft^3.

$$(9{,}156.24\ \text{ft}^3)(7.48\ \text{gal/ft}^3) = 68{,}488.68\ \text{gal, round to } \mathbf{68{,}000\ gal}$$

DETENTION TIME CALCULATIONS

42. Calculate the detention time in hours for the following portion of a treatment plant: Five flocculation basins each 50.1 ft by 18 ft with an average water depth of 11.5 ft. A sedimentation (sed) basin that is 322 ft long, 75 ft wide, and has an average water depth of 10.0 ft. The flow through these basins is 15.6 mgd.

First, determine the number of gallons in the five flocculation basins and the sed basin.

The equation is:

$$\textbf{Volume, gal} = \textbf{(Length)(Width)(Depth)(7.48 gal/ft}^3\textbf{)}(5\ \text{basins})$$

$$\text{Volume, gal in floc basins} = (50.1\ \text{ft})(18\ \text{ft})(11.5\ \text{ft})(7.48\ \text{gal/ft}^3)(5\ \text{basins}) = 387{,}864\ \text{gal}$$

$$\text{Volume, gal in sed basin} = (322\ \text{ft})(75\ \text{ft})(10.0\ \text{ft})(7.48\ \text{gal/ft}^3) = \underline{1{,}806{,}420\ \text{gal}}$$

$$2{,}194{,}284\ \text{gal}$$

Next, convert million gallons per day to gallons per hour.

$$(15.6\ \text{mgd})(1\ \text{d}/24\ \text{hr})(1{,}000{,}000/1\ \text{M}) = 650{,}000\ \text{gph}$$

Write the equation with units asked for in the question.

$$\textbf{Detention time, hr} = \frac{\textbf{Volume, gal}}{\textbf{Flow rate, gph}}$$

$$\text{Detention time, hr} = \frac{2{,}194{,}284\ \text{gal}}{650{,}000\ \text{gph}} = 3.38\ \text{hr, round to } \mathbf{3.4\ hr}$$

Detention times are theoretical because basins begin to fill with settled floc and other debris and the probability that there is some short-circuiting.

43. Find the detention time in hours for a clarifier that has a diameter of 160.0 ft, a water depth of 10.25 ft, and a flow rate of 3.86 mgd.

First, determine the volume in gallons for the clarifier.

$$\textbf{Volume, gal} = \textbf{(0.785)(Diameter)}^2\textbf{(Depth)(7.48 gal/ft}^3\textbf{)}$$

$$\text{Volume, gal} = (0.785)(160.0 \text{ ft})(160.0 \text{ ft})(10.25 \text{ ft})(7.48 \text{ gal/ft}^3) = 1{,}540{,}760 \text{ gal}$$

Then, convert million gallons per day to gallons per hour, because detention time is asked for in hours.

$$(3.86 \text{ mgd})(1{,}000{,}000/1 \text{ M})(1 \text{ d}/24 \text{ hr}) = 160{,}833 \text{ gph}$$

The equation is:

$$\textbf{Detention time, hr} = \frac{\textbf{Volume, gal}}{\textbf{Flow rate, gph}}$$

$$\text{Detention time, hr} = \frac{1{,}540{,}760 \text{ gal}}{160{,}833 \text{ gph}} = \textbf{9.58 hr}$$

PRESSURE PROBLEMS

44. What is the psi at the bottom of a tank if the water level is 24.3 ft deep?

The equation is:

$$\textbf{psi} = \frac{\textbf{Depth, ft}}{\textbf{2.31 ft/psi}}$$

$$\text{psi} = \frac{24.3 \text{ ft}}{2.31 \text{ ft/psi}} = \textbf{10.5 psi}$$

45. What is the depth of water in a lake if the psi is 40.7?

The equation is:

$$\textbf{psi} = \frac{\textbf{Depth, ft}}{\textbf{2.31 ft/psi}}$$

Rearrange and solve.

$$\text{Depth, ft} = (40.7 \text{ psi})(2.31 \text{ ft/psi}) = \textbf{94.0 ft deep}$$

46. If the water depth in a reservoir is 13.06 ft, what is the pressure at 12 ft below the surface in pounds per square foot?

The equation using pressure in pounds per square foot is:

$$\textbf{Pressure, lb/ft}^2 = \textbf{(Depth)(Density, 62.4 lb/ft}^3\textbf{)}$$

$$\text{Pressure} = (12 \text{ ft})(62.4 \text{ lb/ft}^3) = 748.8 \text{ lb/ft}^2\text{, round to } \textbf{750 lb/ft}^2$$

DENSITY CALCULATIONS

The density of a substance is the amount of mass for a given volume.

47. What is the density in pounds per gallon of a 1.000-L solution that weighs 2.29 lb?

First, convert milliliters to gallons.

$$(1.000 \text{ L})(1 \text{ gal}/3.785 \text{ L}) = 0.2642 \text{ gal}$$

The equation is:

$$\textbf{Density} = \textbf{Mass/Volume}$$

$$\text{Density of solution} = 2.29 \text{ lb}/0.2642 \text{ gal} = \textbf{8.67 lb/gal}$$

48. A substance weighs 378 g and occupies a space of 48 cm^3. What is its density in grams per cubic centimeter?

The equation is:

$$\textbf{Density} = \textbf{Mass/Volume}$$

$$\text{Density} = 378 \text{ g}/48 \text{ cm}^3 = \mathbf{7.9\ g/cm^3}$$

49. Find the density of a solution in grams per cubic centimeter if it weighs 9.50 lb/gal.

$$\text{Number of g/cm}^3 = (\text{Number of lb/gal})(454 \text{ g}/1 \text{ lb})(1 \text{ gal}/3{,}785 \text{ cm}^3)$$

$$\text{Number of g/cm}^3 = (9.50 \text{ lb/gal})(454 \text{ g}/1 \text{ lb})(1 \text{ gal}/3{,}785 \text{ cm}^3) = \mathbf{1.14\ g/cm^3}$$

50. The density of an unknown substance is 2.14 g/cm^3. How much space would this substance occupy in cubic centimeters if it weighed 3.3 lb?

First, convert the number of pounds to grams.

$$\text{Number of g} = (\text{Number of lb})(454 \text{ g}/1 \text{ lb})$$

Substituting:

$$\text{Number of g} = (3.3 \text{ lb})(454 \text{ g}/1 \text{ lb}) = 1{,}498.2 \text{ g}$$

We know that 2.15 g of the substance occupies 1 cm^3 by knowing its density. To get the space 1,498.2 g occupies, divide by the density.

$$\text{Space occupied by substance} = \frac{1{,}498.2 \text{ g}}{2.14 \text{ g/cm}^3} = \mathbf{700\ cm^3}$$

51. The density of an unknown substance is 1.87 g/cm^3. How much space would this substance occupy in cubic centimeters if it weighed 25 lb?

First, convert the number of pounds to grams.

$$\text{Number of g} = (\text{Number of lb})(454\ \text{g}/1\ \text{lb})$$

Substituting:

$$\text{Number of g} = (25\ \text{lb})(454\ \text{g}/1\ \text{lb}) = 11{,}350\ \text{g}$$

We know that 1.87 g of the substance occupies 1 cm^3 by knowing its density. To get the space 11,350 g occupies, divide by the density.

$$\text{Space occupied by substance} = \frac{11{,}350\ \text{g}}{1.87\ \text{g/cm}^3} = 6{,}069.5\ \text{cm}^3\text{, round to } \mathbf{6{,}100\ cm^3}$$

52. The specific gravity (sp gr) of a liquid solution is 1.24. How many pounds will 1 ft^3 weigh?

Rearranging the above equation:

$$\text{Pounds per ft}^3 = (\text{Specific gravity})(62.4\ \text{lb/ft}^3)$$

$$\begin{aligned}\text{Pounds per gallon of liquid solution} &= (1.24)(62.4\ \text{lb/ft}^3)\\ &= \mathbf{77.4\ lb\ for\ 1\ ft^3\ of\ the\ liquid\ solution}\end{aligned}$$

SPECIFIC GRAVITY OF LIQUIDS

Specific gravity compares the density of one substance to another. Water is the standard for liquids.

53. The density of an unknown liquid is 56.4 lb/ft^3. What is the specific gravity of the liquid?

Water has a density of 62.4 lb/ft^3. Divide the density of the unknown by the density of water.

The equation is:

Specific gravity = Density of substance/Density of water

$$\text{Specific gravity of unknown substance} = \frac{56.4 \text{ lb/ft}^3}{62.4 \text{ lb/ft}^3} = \mathbf{0.904 \text{ sp gr}}$$

54. What is the specific gravity for a solution that weighs 10.18 lb/gal?

The density of water can also be expressed as pounds per gallon, or 8.34 lb/gal.

$$\text{Specific gravity} = \frac{10.18 \text{ lb/gal}}{8.34 \text{ lb/gal}} = \mathbf{1.22 \text{ sp gr}}$$

FLOW PROBLEMS

55. What is the velocity of flow in feet per second for a 6-in. diameter pipe if it delivers 275 gpm?

First, convert the number of gallons per minute to cubic feet per second.

$$\text{Number of ft}^3\text{/s} = \frac{275 \text{ gpm}}{(7.48 \text{ gal/ft}^3)(60 \text{ s/min})} = 0.613 \text{ ft}^3\text{/s}$$

Next, convert the diameter from inches to feet.

$$\text{Number of ft} = (6 \text{ in.})(1 \text{ ft/12 in.}) = 0.5 \text{ ft}$$

The equation is:

$$\textbf{Flow, ft}^3\textbf{/s} = \textbf{(Area, ft}^2\textbf{)(Velocity, ft/s)}$$

where the area = (0.785)(Diameter)^2

$$0.613 \text{ ft}^3\text{/s} = (0.785)(0.5)(0.5)(\text{Velocity, ft/s})$$

Rearrange and solve for the flow in feet per second.

$$\text{Flow, ft/s} = \frac{0.613 \text{ ft}^3\text{/s}}{(0.785)(0.5)(0.5)} = 3.12 \text{ ft/s, round to } \textbf{3 ft/s}$$

56. Water is flowing through a channel that is 6.0 ft wide and 3.0 ft deep. If the flow is 28.0 ft^3/s, what is the velocity of the water?

The equation is:

$$\textbf{Q (Flow)} = \textbf{(Area)(Velocity)}$$

Substitute known parameters.

$$28.0 \text{ ft}^3\text{/s} = (6.0 \text{ ft})(3.0 \text{ ft})(\text{Velocity})$$

Rearrange and solve for velocity.

$$\text{Velocity} = \frac{28.0 \text{ ft}^3\text{/s}}{(6.0 \text{ ft})(3.0 \text{ ft})} = \textbf{1.6 ft/s}$$

CALCULATIONS OF CHLORINE DOSE, DEMAND, AND RESIDUAL

57. If the chlorine residual is 0.8 mg/L and the chlorine demand is 2.8 mg/L, what is the chlorine dose?

The equation is:

Chlorine dose = Chlorine demand + Chlorine residual

Chlorine dose = 2.8 mg/L + 0.8 mg/L = **3.6 mg/L**

58. If the chlorine dose is 3.95 mg/L and the chlorine demand is 2.43 mg/L, what is the chlorine residual?

The equation is:

Chlorine dose = Chlorine demand + Chlorine residual

Rearrange and solve for chlorine residual.

Chlorine residual = Chlorine dose – Chlorine demand

Chlorine residual = 3.95 mg/L – 2.43 mg/L = **1.52 mg/L**

DOSAGE PROBLEMS

***Note*:** For convenience, most of the dosage problems in this book do not take chlorine demand into account.

59. If the desired dose for chlorine is 2.0 mg/L, how many pounds of chlorine per day will be needed to treat 2.1 mgd?

Use the "pounds per day" formula.

Pounds per day = **(mgd)(Dosage, mg/L)(8.34 lb/gal)**

Substituting:

Pounds per day = (2.1 mgd)(2.0 mg/L)(8.34 lb/gal) = **35 lb/d of Cl_2**

60. A water treatment facility is treating 8.25 mgd. If the desired dose for chlorine is 1.8 mg/L, how many pounds of chlorine per day are required?

The equation is:

Pounds per day = (mgd)(Dosage, mg/L)(8.34 lb/gal)

Pounds per day = (8.25 mgd)(1.8 mg/L)(8.34 lb/gal)

= 123.8 lb/d, round to **120 lb/d of Cl_2**

61. A water treatment facility is treating 5.8 mgd. If the desired dose for chlorine is 1.5 mg/L, how many pounds of chlorine per day are required?

The equation is:

Pounds per day = (mgd)(Dosage, mg/L)(8.34 lb/gal)

Substituting:

Pounds per day = (5.8 mgd)(1.5 mg/L)(8.34 lb/gal)

= 72.6 lb/d, round to **73 lb/d of Cl_2**

62. What is the chlorine dosage in milligrams per liter if 4.6 mgd is treated with 55 lb/d of chlorine?

The equation is:

Pounds per day = (mgd)(Dosage, mg/L)(8.34 lb/gal)

Substituting and rearranging the formula:

$$\frac{(55 \text{ lb/d})}{(4.6 \text{ mgd})(8.34 \text{ lb/gal})} = \text{Dosage}$$

Dosage, mg/L = **1.4 mg/L of Cl_2**

63. How many pounds per day of liquid alum are used if the flow rate is 12.7 mgd and the dosage is 8.25 mg/L? The purity of the alum is 48.0%.

Convert the percent purity to decimal form.

$$48.0\%/100\% = 0.480$$

The equation is:

$$\textbf{Pounds per day} = \frac{\textbf{(mgd)(Dosage, mg/L)(8.34 lb/gal)}}{\textbf{\% purity}}$$

$$\text{Pounds per day, alum} = \frac{(12.7 \text{ mgd})(8.25 \text{ mg/L})(8.34 \text{ lb/gal})}{48.0\%/100\% \text{ purity}}$$

Pounds per day, alum = **1,820 lb/d of alum**

64. How many pounds per day of lime are being used to treat 4.5 mgd with a dosage of 7.2 mg/L? The lime is 89% pure.

The equation is:

$$\textbf{Pounds per day, lime} = \frac{\textbf{(mgd)(Dosage, mg/L)(8.34 lb/gal)}}{\textbf{\% purity}}$$

$$\text{Pounds per day, lime} = \frac{(4.5 \text{ mgd})(7.2 \text{ mg/L})(8.34 \text{ lb/gal})}{89.0\%/100\% \text{ purity}}$$

= 304 lb/d, round to **300 lb/d of lime**

65. A treatment plant is using 880 lb/d of chlorine gas. If the chlorine demand is 2.8 mg/L and the chlorine residual is 1.5 mg/L, how many million gallons per day are being treated?

First, find the total chlorine dosage.

$$\textbf{Total } \mathbf{Cl_2} \textbf{ dosage} = \mathbf{Cl_2} \textbf{ Demand} + \mathbf{Cl_2} \textbf{ Residual}$$

$$\text{Total } Cl_2 \text{ dosage} = 2.8 \text{ mg/L (Demand)} + 1.5 \text{ mg/L (Residual)} = 4.3 \text{ mg/L}$$

Then, use the "pounds" formula, but solve for the unknown, million gallons per day.

$$\text{Pounds per day} = (\text{mgd})(\text{Dosage})(8.34 \text{ lb/gal})$$

$$\text{Million gallons per day} = \frac{\text{lb/d}}{(\text{Dosage})(8.34 \text{ lb/gal})}$$

$$\text{Million gallons per day} = \frac{880 \text{ lb/d}}{(4.3 \text{ mg/L})(8.34 \text{ lb/gal})} = 24.5 \text{ mgd, round to } \mathbf{25 \text{ mgd}}$$

66. A plant is treating water at 72.3 mgd. If lime is being added at a rate of 1,645.94 g/min, what is the lime usage in pounds per day and the dosage in milligrams per liter?

First, find the pounds per day of lime usage.

$$\text{Pounds per day of lime} = (\text{g/min})(1{,}440 \text{ min/d})(1 \text{ lb/454 g}) = \text{lb/d}$$

$$\text{Pounds per day of lime} = (1{,}645.94 \text{ g/min})(1{,}440 \text{ min/d})(1 \text{ lb/454 g})$$

$$= \mathbf{5{,}220.60 \text{ lb/d of lime}}$$

Then, using the "pounds" equation, calculate the dosage in milligrams per liter by rearranging the formula and solving for dosage.

$$\textbf{Dosage, mg/L} = \frac{\textbf{lb/day}}{\textbf{(mdg)(8.34 lb/gal)}}$$

$$\text{Dosage, mg/L} = \frac{5{,}220.60 \text{ lb/d}}{(72.3 \text{ mgd})(8.34 \text{ lb/gal})} = \mathbf{8.66 \text{ mg/L of lime}}$$

67. A treatment plant is treating 24.2 mgd with a 64% calcium hypochlorite solution. The dosage is 1.25 mg/L. If the flow rate is reduced to 17.8 mgd and the dosage is also reduced to 1.20 mg/L, how many pounds per day of sodium hypochlorite are needed for the new flow rate?

The equation is:

$$\textbf{Pounds per day} = \frac{\textbf{(mgd)(Dosage, mg/L)(8.34 lb/gal)}}{\textbf{\% purity}}$$

Notice that you were given some data that was not required for answering the problem.

Substituting:

$$\text{Pounds per day of calcium hypochlorite} = \frac{(17.8 \text{ mgd})(1.20 \text{ mg/L})(8.34 \text{ lb/gal})}{64\%/100\%}$$

$$\text{Pounds per day of calcium hypochlorite} = \frac{(183.1 \text{ lb/d})}{64\%/100\%}$$

Pounds per day of calcium hypochlorite =

278.3 lb/d, round to **278 lb/d calcium hypochlorite**

68. Water from a well is being treated with a 5.0% sodium hypochlorite solution. The hypochlorinator is pumping at a rate of 33.6 gpd. What is the chlorine dosage in milligrams per liter if the well is producing 325 gpm? Assume the hypochlorite solution is 8.34 lb/gal.

First, convert the production rate of the pump to million gallons per day.

$$\text{Million gallons per day} = \frac{(\text{Pumping rate})(1{,}440 \text{ min/d})}{1{,}000{,}000/\text{M}}$$

$$\text{Million gallons per day} = \frac{(325 \text{ gpm})(1{,}440 \text{ min/d})}{1{,}000{,}000/\text{M}} = 0.468 \text{ mgd}$$

Next, calculate the chlorine usage in pounds per day.

$$\textbf{Chlorine usage, lb/d} = \frac{\textbf{(Hypochlorinator flow, gpd)(\% hypochlorite)(8.34 lb/gal)}}{\textbf{100\%}}$$

$$\text{Chlorine usage, lb/d} = \frac{(33.6 \text{ gpd})(5.0\%)(8.34 \text{ lb/gal})}{100\%} = 14.01 \text{ lb/d, round to } \mathbf{14 \text{ lb/d}}$$

Last, calculate the chlorine dosage in milligrams per liter using the "pounds" formula.

$$\textbf{Chlorine dosage, mg/L} = \frac{\textbf{Chlorine, lb/d}}{\textbf{(Well flow, mgd)(8.34 lb/gal)}}$$

$$\text{Chlorine dosage, mg/L} = \frac{14.01 \text{ lb/d}}{(0.468 \text{ mgd})(8.34 \text{ lb/gal})} = \mathbf{3.6\ mg/L}$$

69. A 5-min drawdown test result showed that 120 mL of a polymer aid was being used to treat the raw water. The specific gravity of the polymer is 1.18. If the plant is treating 24 ft^3/s, what is the polymer dosage in milligrams per liter?

First, determine the number of milliliters per minute for the polymer.

$$\frac{(120 \text{ mL})}{5 \text{ min}} = 24 \text{ mL/min}$$

Next, calculate the pounds per gallon for the polymer.

$$\text{Pounds per gallon} = (\text{Specific gravity})(8.34 \text{ lb/gal})$$

$$\text{Pounds per gallon, polymer} = (1.18)(8.34 \text{ lb/gal}) = 9.84 \text{ lb/gal}$$

Then, find the number of million gallons per day.

$$(24 \text{ ft}^3\text{/s})(86{,}400 \text{ s/d})(7.48 \text{ gal/ft}^3)(1 \text{ M}/1{,}000{,}000) = 15.5 \text{ mgd}$$

Use the dosage equation with conversions added for convenience (dosage/conversion equation).

$$\textbf{Milligrams per liter, polymer} = \frac{\textbf{(mL/min)(1,440 min/d)(lb/gal, polymer)}}{\textbf{(3,785 mL/gal)(mgd)(8.34 lb/gal)}}$$

$$\text{Milligrams per liter, polymer} = \frac{(24 \text{ mL/min})(1{,}440 \text{ min/d})(9.84 \text{ lb/gal})}{(3{,}785 \text{ mL/gal})(15.5 \text{ mgd})(8.34 \text{ lb/gal})}$$

$$= \mathbf{0.70\ mg/L\ of\ polymer\ aid}$$

70. A small tank containing 775 gal of water is to be disinfected using a sodium hypochlorite solution. If the dosage is 50.0 mg/L and the available chlorine in the solution is 5.0%, how much hypochlorite solution should be added, in ounces? Assume the sodium hypochlorite (hypo) weighs 8.92 lb/gal.

First, solve the problem using the modified "pounds" formula.

The equation is:

$$\textbf{Pounds} = \frac{\textbf{(mil gal)(Dosage, mg/L)(8.34 lb/gal)}}{\textbf{\% purity}}$$

Delete "day" on each side of the equation, because it is not what we're looking for.

Next, find how many million gallons there are in 775 gal.

$$\text{Million gallons} = \frac{775 \text{ gal}}{1{,}000{,}000/\text{M}} = 0.000775 \text{ mil gal}$$

Substituting:

$$\text{Pounds of hypo} = \frac{(0.000775 \text{ mil gal})(50.0 \text{ mg/L})(8.34 \text{ lb/gal})}{5\%/100\%} = 6.46 \text{ lb}$$

Now, convert pounds to gallons.

$$\frac{6.46 \text{ lb of hypo}}{8.92 \text{ lb/gal}} = 0.72 \text{ gal}$$

Now, convert gallons to ounces.

Number of ounces, hypo = (128 oz/gal)(0.72 gal)

= 92.16 oz, round to **92 oz of hypochlorite**

71. A 325,000-gal water tank is to be disinfected with a 61% hypochlorite solution. The dosage desired is 50.0 mg/L. How many liters of hypochlorite will be required?

First, convert 325,000 gal to million gallons.

$$325{,}000 \text{ gal} = \frac{0.325 \text{ mil gal}}{1{,}000{,}000/\text{M}}$$

Next, determine the number of pounds of hypochlorite required.

The equation is:

$$\textbf{Pounds} = \frac{\textbf{(mgd)(Dosage, mg/L)(8.34 lb/gal)}}{\textbf{\% purity}}$$

Delete "day" on each side of the equation because it is not needed.

$$\text{Pounds, hypo} = \frac{(0.325 \text{ mil gal})(50.0 \text{ mg/L})(8.34 \text{ lb/gal})}{61\%/100\%} = 222.17 \text{ lb of hypo}$$

Now, convert the number of pounds of hypo to gallons.

$$\frac{222.17 \text{ lb of hypo}}{8.34 \text{ lb/gal}} = 26.639 \text{ gal}$$

Last, convert gallons to liters.

(26.639 gal)(3.785 L/gal) = 100.829 L of hypo, round to **100 L of hypochlorite**

***Note*: The 8.34 lb/gal multiplier in the first step can be dropped in the "pounds" equation. However, to avoid confusion, it was retained here, thus the need for the second step.**

72. How many milliliters per minute of cationic polymer are needed if the desired dose is 1.40 mg/L and the plant is treating 37 ft^3/s? The polymer has a specific gravity of 1.20 and is 89.8% pure.

First, convert cubic feet per second to million gallons per day.

$$\frac{(37 \text{ ft}^3\text{/s})(86{,}400 \text{ s/d})(7.48 \text{ gal/ft}^3)}{1{,}000{,}000/1 \text{ mil gal}} = 23.91 \text{ mgd}$$

Next, find the number of pounds per gallon for the polymer.

$$\frac{(1.20)(8.34 \text{ lb/gal})(89.8\% \text{ purity})}{100\%} = 8.99 \text{ lb/gal}$$

Next, find the number of pounds per day of polymer using the "pounds" formula.

Pounds per day = (mgd)(Dosage, mg/L)(8.34 lb/gal)

Pounds per day = (23.91 mgd)(1.40 mg/L)(8.34 lb/gal)

Pounds per day = 279.17 lb/d

(continued)

Then, divide the number of pounds per day by the pounds per gallon and number of minutes in a day to get gallons per minute.

$$\frac{(279.17 \text{ lb/d})}{(8.99 \text{ lb/gal})(1{,}440 \text{ min/d})} = 0.0216 \text{ gal/min}$$

Last, convert gallons per minute to milliliters per minute.

$$(3{,}785 \text{ mL/gal})(0.0216 \text{ gal/min}) = 81.76 \text{ mL/min, round to } \mathbf{82 \text{ mL/min}}$$

73. A 5-min drawdown test result shows that 262 mL of alum is being injected. The alum amount is 10.46 lb/gal. If the plant is treating 925 gpm, what is the alum dosage in milligrams per liter?

First, determine the number of milliliters per minute for the alum.

$$\frac{(262 \text{ mL})}{5 \text{ min}} = 52.4 \text{ mL/min}$$

Next, find the number of million gallons per day being treated.

$$(925 \text{ gpm})(1{,}440 \text{ min/d})(1 \text{ M}/1{,}000{,}000) = 1.332 \text{ mgd}$$

The dosage/conversion equation is:

$$\textbf{Milligrams per liter, alum} = \frac{\textbf{(mL/min)(1,440 min/d)(lb/gal, alum)}}{\textbf{(3,785 mL/gal)(mgd)(3.84 lb/gal)}}$$

$$\text{Milligrams per liter, alum} = \frac{(52.4 \text{ mL/min})(1{,}440 \text{ min/d})(10.46 \text{ lb/gal})}{(3{,}785 \text{ mL/gal})(1.332 \text{ mgd})(8.34 \text{ lb/gal})}$$

$$= \mathbf{18.8 \text{ mg/L of alum}}$$

74. Determine the feed rate for alum in milliliters per minute under the following conditions:

Plant flow	**75 mgd**	**Alum percentage**	**48%**
Alum dosage rate	**11.2 mg/L**	**Alum specific gravity**	**1.22**

First, find the pounds per gallon for alum.

$$\text{Pounds per gallon alum} = (\text{Specific gravity})(8.34\ \text{lb/gal})$$

$$\text{Pounds per gallon alum} = (1.22)(8.34\ \text{lb/gal}) = 10.17\ \text{lb/gal}$$

The equation for dosage is:

$$\textbf{Dosage, mg/L} = \frac{\textbf{(mL/min)(1,440 min/d)(Alum, lb/gal)(\% purity)}}{\textbf{(mgd)(8.34 lb/gal)(3,785 mL/gal)}}$$

Rearrange to solve for the unknown (milliliters per minute).

$$\text{Alum, mL/min} = \frac{(\text{Dosage, mg/L})(\text{mgd})(8.34\ \text{lb/gal})(3{,}785\ \text{mL/gal})}{(\text{Alum, lb/gal})(\%\ \text{purity})(1{,}440\ \text{min/d})}$$

$$\text{Alum, mL/min} = \frac{(11.2\ \text{mg/L})(75\ \text{mgd})(8.34\ \text{lb/gal})(3{,}785\ \text{mL/gal})}{(10.17\ \text{lb/gal})(48\%/100\%\ \text{purity})(1{,}440\ \text{min/d})} = \textbf{3,800 mL/min of alum}$$

75. Calculate the feed rate for fluorosilicic acid in milliliters per minute given the following data:

Flow rate is 53.5 mgd	**Treated with 24% solution of H_2SiF_6**
Fluoride desired is 1.1 mg/L	**Fluoride ion percent is 79.1%**
Fluoride in raw water is 0.15 mg/L	**H_2SiF_6 weighs 9.86 lb/gal**

First, determine the required fluoride dose.

$$\text{F req.} = \text{F desired} - \text{F in raw water}$$

$$\text{F req.} = 1.1\ \text{mg/L} - 0.15\ \text{mg/L} = 0.95\ \text{mg/L}$$

Next, calculate the milligrams per liter of H_2SiF_6.

$$\text{Number of mg/L } H_2SiF_6 = \frac{(\%\ \text{solution})(10{,}000\ \text{mg/L})}{1\%} = \frac{(24\%\ \text{F})(10{,}000\ \text{mg/L})}{1\%}$$

$$\text{Number of mg/L } H_2SiF_6 = 240{,}000\ \text{mg/L}$$

(continued)

Now, solve for the million gallons per day flow of fluoride.

The equation is:

(F, mg/L)(mgd)(8.34 lb/gal) = (% as mg/L)(mgd flow)(Solution, lb/gal)(% F ion)

The following applies:

$$\%F = 79.1\%/100\% = 0.791 \text{ F}$$

$$(0.95 \text{ mg/L F})(53.5 \text{ mgd})(8.34 \text{ lb/gal}) = (240{,}000 \text{ mg/L})(x \text{ mgd flow})(9.86 \text{ lb/gal})(0.791 \text{ F})$$

Solve for x.

$$x \text{ mgd flow} = \frac{(0.95 \text{ mg/L F})(53.5 \text{ mgd})(8.34 \text{ lb/gal})}{(240{,}000 \text{ mg/L})(9.86 \text{ lb/gal})(0.791 \text{ F})} = 0.000226 \text{ mgd}$$

Convert million gallons per day to gallons.

$$(0.000226 \text{ mgd})(1{,}000{,}000 \text{ gal/1 M}) = 226 \text{ gpd } H_2SiF_6$$

Then, convert gallons per day to milliliters per minute.

$$\text{Milliliters per minute of } H_2SiF_6 = (226 \text{ gpd})(3{,}785 \text{ mL/gal})(1 \text{ d/1,440 min})$$

$$= 594 \text{ mL/min, round to } \mathbf{590 \text{ mL/min of } H_2SiF_6}$$

76. A plant treats 8.5 mgd with alum and a coagulant aid that weighs 10.27 lb/gal. The results of a drawdown test are 98 mL for the coagulant aid in 5 min. What is the coagulant aid dosage in milligrams per liter?

First, determine the number of milliliters per minute for the coagulant aid.

$$\frac{98 \text{ mL}}{5 \text{ min}} = 19.6 \text{ mL/min}$$

The dosage equation is:

$$\mathbf{Milligrams\ per\ liter,\ coagulant\ aid} = \frac{\mathbf{(mL/min)(1{,}440\ min/d)(lb/gal,\ coagulant\ aid)}}{\mathbf{(3{,}785\ mL/gal)(mil\ gal)(8.34\ lb/gal)}}$$

$$\text{Milligrams per liter, coagulant aid} = \frac{(19.6 \text{ mL/min})(1{,}440 \text{ min/d})(10.27 \text{ lb/gal})}{(3{,}785 \text{ mL/gal})(8.5 \text{ mil gal})(8.34 \text{ lb/gal})}$$

$$\text{Milligrams per liter, coagulant aid} = \mathbf{1.1 \text{ mg/L of coagulant aid}}$$

77. A water treatment plant has a filter effluent flow of 4,400 gpm and is being treated with 855 gpd of a hypochlorite solution. The desired dose is 2.5 mg/L; determine the concentration of the hypochlorite solution in percent.

First, convert 4,400 gpm filter flow to million gallons per day.

$$\frac{(4{,}400 \text{ gpm})\ (1{,}440 \text{ min})\quad (1 \text{ mil gal})}{\text{day} \qquad 1{,}000{,}000 \text{ gal}} = 6.336 \text{ mgd}$$

Then, convert 855 gpd of the hypochlorite solution rate of flow to million gallons per day.

$$\frac{(855 \text{ gpd})\quad (1 \text{ mil gal})}{1{,}000{,}000 \text{ gal}} = 0.000855 \text{ mgd}$$

Then, solve by using the equal dosage equations.

$$(0.000855 \text{ mgd})(x \text{ mg/L})(8.34 \text{ lb/gal}) = (6.336 \text{ mgd})(2.5 \text{ mg/L})(8.34 \text{ lb/gal})$$

$$x \text{ mg/L} = \frac{(6.336 \text{ mgd})(2.5 \text{ mg/L})(8.34 \text{ lb/gal})}{(0.000855 \text{ mgd})(8.34 \text{ lb/gal})} = 18{,}526 \text{ mg/L}$$

Last, convert milligrams per liter to percent.

$$\frac{(18{,}526 \text{ mg/L})\qquad (1\%)}{10{,}000 \text{ mg/L}} = \mathbf{1.9\%}$$

78. A water treatment plant has a filter effluent flow of 3,860 gpm and is being treated with 792 gpd of a hypochlorite solution. The desired dose is 1.75 mg/L; determine the percent concentration of the hypochlorite solution (assume 8.34 lb/gal).

First, convert 3,860 gpm filter flow to million gallons per day.

$$\frac{(3{,}860 \text{ gpm})\quad (1{,}440 \text{ min})\quad (1 \text{ mil gal})}{\text{day} \qquad 1{,}000{,}000 \text{ gal}} = 5.558 \text{ mgd}$$

Then, convert 792 gpd of the hypochlorite solution rate of flow to million gallons per day.

$$\frac{(792 \text{ gpd})\quad (1 \text{ mil gal})}{1{,}000{,}000 \text{ gal}} = 0.000792 \text{ mgd}$$

(continued)

Then, using the equal dosage equations:

$$(0.000792 \text{ mgd})(x \text{ mg/L})(8.34 \text{ lb/gal}) = (5.558 \text{ mgd})(1.75 \text{ mg/L})(8.34 \text{ lb/gal})$$

$$x \text{ mg/L} = \frac{(5.558 \text{ mgd})(1.75 \text{ mg/L})(8.34 \text{ lb/gal})}{(0.000792 \text{ mgd})(8.34 \text{ lb/gal})} = 12{,}281 \text{ mg/L}$$

Last, convert milligrams per liter to percent.

$$\frac{(12{,}281 \text{ mg/L}) \qquad (1\%)}{10{,}000 \text{ mg/L}} = \mathbf{1.23\%}$$

POUNDS REMOVAL PROBLEMS

79. Find the amount of iron (Fe) and manganese (Mn) removed per year from a plant that treats an average of 19.6 mgd if the average iron concentration is 1.26 ppm and the average manganese concentration is 0.17 ppm. The removal efficiency is 82.5% for the iron and 70.3% for the manganese.

First, calculate the iron and manganese removal in ppm.

$$(1.26 \text{ ppm})(82.5\%/100\%) = (1.26 \text{ ppm})(0.825) = 1.04 \text{ ppm}$$

$$(0.17 \text{ ppm})(70.3\%/100\%) = (0.17 \text{ ppm})(0.703) = 0.12 \text{ ppm}$$

Determine the amount of water in million gallons produced for the year.

$$(19.6 \text{ mgd})(365 \text{ d/yr}) = 7{,}154 \text{ mil gal/yr}$$

Next, using the "pounds" equation, solve for the number of pounds per year for iron and manganese.

$$\textbf{Pounds per year} = \textbf{(mil gal/yr)(Dosage, mg/L)(8.34 lb/gal)}$$

$$\text{Pounds per year, Fe} = (7{,}154 \text{ mil gal/yr})(1.040 \text{ mg/L})(8.34 \text{ lb/gal})$$

$$= 62{,}051 \text{ lb/yr, round to } \mathbf{62{,}100 \text{ lb/yr of Fe removed}}$$

$$\text{Pounds per year, Mn} = (7{,}154 \text{ mil gal/yr})(0.120 \text{ mg/L})(8.34 \text{ lb/gal})$$

$$= 7{,}160 \text{ lb/yr, round to } \mathbf{7{,}200 \text{ lb/yr of Mn removed}}$$

80. Find the amount of iron and manganese removed per year from a plant that treats an average of 33.8 mgd if the average iron concentration is 1.29 ppm and the average manganese concentration is 0.27 ppm. The removal efficiency is 92.4% for the iron and 76.4% for the manganese.

First, calculate the iron and manganese removal in parts per million.

$$(1.29 \text{ ppm})(92.4\%/100\%) = 1.29 \text{ ppm}(0.924) = 1.192 \text{ ppm}$$

$$(0.27 \text{ ppm})(76.4\%/100\%) = 0.27 \text{ ppm}(0.764) = 0.206 \text{ ppm}$$

Determine the amount of water in million gallons produced for the year.

$$(33.8 \text{ mgd})(365 \text{ d/yr}) = 12{,}337 \text{ mil gal/yr}$$

Next, using the "pounds" equation, solve for the number of pounds per year for iron and manganese.

$$\textbf{Pounds per year} = \textbf{(mil gal/yr)(Dosage, mg/L)(8.34 lb/gal)}$$

$$\text{Pounds per year, Fe} = (12{,}337 \text{ mil gal/yr})(1.192 \text{ mg/L})(8.34 \text{ lb/gal})$$

$$= 122{,}646 \text{ lb/yr, round to } \textbf{122,000 lb/yr of Fe removed}$$

$$\text{Pounds per year, Mn} = (12{,}337 \text{ mil gal/yr})(0.206 \text{ mg/L})(8.34 \text{ lb/gal})$$

$$= 21{,}195 \text{ lb/yr, round to } \textbf{21,000 lb/yr of Mn removed}$$

PUMP DISCHARGE PROBLEMS

81. If a pump discharges 8,750 gal in 2 hr and 45 min, how many gallons per minute is the pump discharging?

First, determine the number of minutes the pump was working.

$$2 \text{ hr}(60 \text{ min/hr}) + 45 \text{ min} = 120 \text{ min} + 45 \text{ min} = 165 \text{ min}$$

Then, determine the number of gal/min by dividing the number of gallons pumped by the total time the pump worked.

$$\frac{8{,}750 \text{ gal}}{165 \text{ min}} = \textbf{53.0 gal/min}$$

82. How long will it take, in hours, for a pump to discharge 86,400 gal if it is pumping at a rate of 30 gpm?

First, determine the number of minutes it will take.

$$\frac{(86{,}400 \text{ gal})}{(30 \text{ gal/min})} = 2{,}880 \text{ min}$$

Then, convert to hours by dividing by 60 min per hour.

$$\frac{2{,}880 \text{ min}}{60 \text{ min/hr}} = \mathbf{48\ hr}$$

83. A pump's output is averaging 36 gpm. How many gallons will it pump in one day?

(36 gal/min)(1,440 min/d) = 51,840 gal, round to **52,000 gal**

84. How many gallons will a pump discharge if it pumps an average of 65 gpm for 1 hr and 42 min?

First, calculate the number of minutes for the time interval given.

1 hr(60 min/hr) + 42 min = 60 min + 42 min = 102 min

Then, calculate the number of gallons by multiplying the length of time the pump worked in minutes times the number of gallons per minute.

102 min(65 gal/min) = 6,630 gal, round to **6,600 gal**

85. Determine a pump's total output in million gallons per day if it is pumping 1,550 gal/min.

(1,550 gal/min)(1,440 min/d)(1 mil gal/1,000,000) = **2.23 mgd**

PUMPING CALCULATIONS

86. Find the total head in feet for a pump with a total static head of 22 ft and a head loss of 2 ft.

The equation is:

Total head, ft = Total static head, ft = Head losses, ft

Total head, ft = 22 ft + 2 ft = **24 ft**

87. What is the motor horsepower (mhp) if 40 hp is required to run a pump with a motor efficiency of 92% and a pump efficiency of 83%? Note: The 40 hp in this problem is called the water horsepower (whp). The water horsepower is the actual energy (horsepower) available to pump water.

The equation is:

$$\textbf{Motor horsepower} = \frac{\textbf{(whp)}}{\textbf{(Motor efficiency)(Pump efficiency)}}$$

$$\text{Motor horsepower} = \frac{(40\text{ whp})}{(92\%/100\%\text{ Motor efficiency})(83\%/100\%\text{ Pump efficiency})}$$

$$\text{Motor horsepower} = \frac{(40\text{ whp})}{(0.92\text{ Motor efficiency})(0.83\text{ Pump efficiency})}$$

Motor horsepower = 52.4 mhp, round to **50 mhp**

88. Water is being pumped from a water source with an elevation of 290 ft to an elevation of 365 ft. What is the total head if friction and minor head losses are 12 ft?

The equation is:

Total head, ft = Total static head, ft (difference in elevation) + Head losses, ft

Total head, ft = (365 ft – 290 ft) + 12 ft = 75 ft + 12 ft = **87 ft**

89. What is the motor horsepower if the brake horsepower (bhp) is 48 and the motor efficiency is 88%?

The equation is:

Motor horsepower = Brake horsepower/Motor efficiency

Motor horsepower = 48 bhp/(88%/100% Motor efficiency) = **54 mhp**

90. What is the brake horsepower if 10 hp is supplied to a motor with 88% efficiency?

The equation is:

Brake horsepower = (hp)(Motor efficiency)

Brake horsepower = (10 hp)(88%/100% Motor efficiency) = **9 bhp**

91. Find the water horsepower if the brake horsepower is 45.7 and the pump efficiency is 81%.

The equation is:

Water horsepower = (bhp)(Pump efficiency)

Water horsepower = (45.7)(81%/100% Pump efficiency) = **37 whp**

92. What is the brake horsepower if the water horsepower is 40 and the pump efficiency is 78%?

The equation is:

Brake horsepower = Water horsepower/Pump efficiency

Brake horsepower = 40 whp/(78%/100% Pump efficiency)

= 51.3 bhp, round to **50 bhp**

93. What is the motor horsepower if 60 hp (hp = water horsepower = whp) is required to run a pump with a motor efficiency of 93% and a pump efficiency of 85%?

The equation is:

$$\textbf{Motor horsepower} = \frac{\textbf{(whp)}}{\textbf{(Motor efficiency)(Pump efficiency)}}$$

$$\text{Motor horsepower} = \frac{(60 \text{ whp})}{(93\%/100\% \text{ Motor efficiency})(85\%/100\% \text{ Pump efficiency})}$$

$$\text{Motor horsepower} = \frac{(60 \text{ whp})}{(0.93 \text{ Motor efficiency})(0.85 \text{ Pump efficiency})}$$

Motor horsepower = 75.9 mhp, round to **80 mhp**

94. Find the water horsepower if the brake horsepower is 38.4 and the pump efficiency is 84%.

The equation is:

Water horsepower = (bhp)(Pump efficiency)

Water horsepower = (38.4)(84%/100% Pump efficiency) = 32.3 whp, round to **32 whp**

FLOW RATE PROBLEMS

95. What is the flow in gallons per minute from a faucet that fills a 5.0-gal container in 3 min and 6 s?

First, convert the number of seconds to minutes.

$$\frac{6 \text{ s}}{60 \text{ s/min}} = 0.1 \text{ min}$$

The total time is 3.1 min.

The equation is:

Flow = Volume/Time

Flow = 5.0 gal/3.1 min = **1.6 gpm**

96. Water is flowing through a channel that is 3.2 ft wide and 2.2 ft deep. If the flow is 7 ft^3/s, what is the velocity of the water?

The equation is:

Q (Flow) = (Area)(Velocity)

$7 \text{ ft}^3\text{/s} = (3.2 \text{ ft})(2.2 \text{ ft})(\text{Velocity})$

Now, solve for velocity.

$$\text{Velocity} = \frac{7 \text{ ft}^3\text{/s}}{(3.2 \text{ ft})(2.2 \text{ ft})} = 0.99 \text{ ft/s, round to } \mathbf{1 \text{ ft/s}}$$

97. A channel that is 8.3 ft wide has water flowing through it at a depth of 4.5 ft and a velocity of 3.1 ft/s. Find the flow through the channel in ft^3/s.

The equation is:

Q (Flow) = (Area)(Velocity)

$\text{Q, ft}^3\text{/s} = (8.3 \text{ ft})(4.5 \text{ ft})(3.1 \text{ ft/s}) = 115.785 \text{ ft}^3\text{/s, round to } \mathbf{120 \text{ ft}^3\text{/s}}$

WEIR OVERFLOW RATE PROBLEMS

98. A rectangular clarifier has a weir length of 250 ft. What is the weir overflow rate in gallons per day per foot if the flow is 3.5 mgd?

The equation is:

$$\textbf{Weir overflow rate} = \frac{\textbf{Flow, gpd}}{\textbf{Weir length, ft}}$$

First, change 3.5 mgd to gallons per day.

$$3.5 \text{ mgd}(1{,}000{,}000/1 \text{ M}) = 3{,}500{,}000 \text{ gpd}$$

$$\text{Weir overflow rate} = \frac{3{,}500{,}000 \text{ gpd}}{250 \text{ ft}} = \mathbf{14{,}000 \text{ gpd/ft}}$$

99. A rectangular clarifier has a weir length of 250 ft. What is the weir overflow rate in gallons per day per foot if the flow is 7.7 mgd?

The equation is:

$$\textbf{Weir overflow rate} = \frac{\textbf{Flow, gpd}}{\textbf{Weir length, ft}}$$

Change 7.7 mgd to gallons per day.

$$(7.7 \text{ mgd})(1{,}000{,}000/1 \text{ M}) = 7{,}700{,}000 \text{ gpd}$$

$$\text{Weir overflow rate} = \frac{7{,}700{,}000 \text{ gpd}}{250 \text{ ft}} = 30{,}800 \text{ gpd/ft, round to } \mathbf{31{,}000 \text{ gpd/ft}}$$

100. A circular clarifier has a weir length of 120 ft (measured to the nearest foot). What is the weir overflow rate in gallons per day per foot if the flow is 2.34 mgd?

The equation is:

$$\textbf{Weir overflow rate} = \frac{\textbf{Flow, gpd}}{\textbf{Weir length, ft}}$$

Change 2.34 mgd to gallons per day.

$$\textbf{2.34 mgd(1,000,000/1 M) = 2,340,000 gpd}$$

$$\text{Weir overflow rate} = \frac{2{,}340{,}000 \text{ gpd}}{120 \text{ ft}} = \textbf{19,500 gpd/ft}$$

SURFACE LOADING RATE CALCULATIONS

101. What is the surface loading rate for a sedimentation basin that is 249.5 ft by 58 ft if it is treating an instantaneous flow rate of 33 ft^3/s?

First, convert the number of cubic feet per second to gallons per day.

$$(33 \text{ ft}^3/\text{s})(86{,}400 \text{ s/d})(7.48 \text{ gal/ft}^3) = 21{,}326{,}976 \text{ gpd}$$

The equation for surface loading rate is:

$$\textbf{Surface loading rate} = \frac{\textbf{(gpd)}}{\textbf{Number of ft}^2}$$

$$\text{Surface loading rate} = \frac{21{,}326{,}976 \text{ gpd}}{(249.5 \text{ ft})(58 \text{ ft})} = 1{,}473.8 \text{ gpd/ft}^2\text{, round to } \mathbf{1{,}500 \text{ gpd/ft}^2}$$

102. What is the surface loading rate for a sedimentation basin that is 385 ft by 74 ft if it is treating an instantaneous flow rate of 19.0 ft³/s?

First, convert the number of cubic feet per second to gallons per day.

$$(19.0 \text{ ft}^3/\text{s})(86{,}400 \text{ s/d})(7.48 \text{ gal/ft}^3) = 12{,}279{,}168 \text{ gpd}$$

The equation for surface loading rate is:

$$\textbf{Surface loading rate} = \frac{\textbf{(gpd)}}{\textbf{Number of ft}^2}$$

$$\text{Surface loading rate} = \frac{12{,}279{,}168 \text{ gpd}}{(385 \text{ ft})(74 \text{ ft})} = 430.999 \text{ gpd/ft}^2\text{, round to } \mathbf{430 \text{ gpd/ft}^2}$$

FILTRATION RATE PROBLEMS

103. A filter has a surface area of 450 ft. What is the filtration rate in gallons per minute if it receives a flow of 1,500 gpm?

The equation is:

$$\textbf{Filtration rate} = \frac{\textbf{Flow rate, gpm}}{\textbf{Filter surface area, ft}^2}$$

$$\text{Filtration rate} = \frac{1{,}500 \text{ gpm}}{450 \text{ ft}^2} = \mathbf{3.3 \text{ gpm/ft}^2}$$

104. A filter is 24 ft by 32 ft. What is the filtration rate in gallons per minute if it receives a flow of 4,050 gpm?

First, find the surface area of the filter.

$$\text{Filter surface area} = (24 \text{ ft})(32 \text{ ft}) = 768 \text{ ft}^2$$

The equation is:

$$\textbf{Filtration rate} = \frac{\textbf{Flow rate, gpm}}{\textbf{Filter surface area, ft}^2}$$

$$\text{Filtration rate} = \frac{4{,}050 \text{ gpm}}{768 \text{ ft}^2} = 5.27 \text{ gpm/ft}^2\text{, round to } \mathbf{5.3 \text{ gpm/ft}^2}$$

UNIT FILTER RUN VOLUME DETERMINATIONS

105. A filter produced a total of 2,152,400 gal between backwashes. If the filter is 25 ft by 20 ft (measured to nearest foot), what is the unit filter run volume (UFRV) in gallons per square foot?

The equation is:

$$\textbf{UFRV, gal/ft}^2 = \frac{\textbf{Total gallons filtered}}{\textbf{Filter surface area, ft}^2}$$

$$\text{UFRV, gal/ft}^2 = \frac{2{,}152{,}400 \text{ gal}}{(25 \text{ ft})(20 \text{ ft})} = 4{,}305 \text{ gal/ft}^2\text{, round to } \mathbf{4{,}300\ gal/ft^2}$$

106. A filter measures 22 ft by 30 ft (measured to nearest foot). Find the UFRV in gallons per square foot if, between backwashes, the total volume of water filtered was 1,202,000 gal.

The equation is:

$$\textbf{UFRV, gal/ft}^2 = \frac{\textbf{Total gallons filtered}}{\textbf{Filter surface area, ft}^2}$$

$$\text{UFRV} = \frac{1{,}202{,}000 \text{ gal}}{(22 \text{ ft})(30 \text{ ft})} = 1{,}821 \text{ gal/ft}^2\text{, round to } \mathbf{1{,}800\ gal/ft^2}$$

107. The UFRV for a filter has been averaging 7,238 gal/ft² over the last 5 years. The filter measures 25 ft by 30 ft (measured to nearest foot). The last filter run between backwashes was 4,075,000 gal. Has this last filter run been above or below average?

The equation is:

$$\textbf{UFRV, gal/ft}^2 = \frac{\textbf{Total gallons filtered}}{\textbf{Filter surface area, ft}^2}$$

$$\text{UFRV} = \frac{4{,}075{,}000 \text{ gal}}{(25 \text{ ft})(30 \text{ ft})} = \mathbf{5{,}400\ gal/ft^2}$$

This is below average.

CALCULATIONS FOR WELLS

108. A well produces 125 gpm. If the drawdown for the well is 20 ft, find the specific yield in gallons per minute per foot.

The equation is:

$$\textbf{Specific yield, gpm/ft} = \frac{\textbf{Well yield, gpm}}{\textbf{Drawdown, ft}}$$

$$\text{Specific yield, gpm/ft} = \frac{125\text{ gpm}}{20\text{ ft}} = 6.25\text{ gpm/ft, round to }\mathbf{6\text{ gpm/ft}}$$

109. Find the drawdown of a well if the well yields 215 gpm and the specific yield is 19.6.

Write the equation, arranging it to solve for the unknown, drawdown.

$$\textbf{Drawdown, ft} = \frac{\textbf{Well yield, gpm}}{\textbf{Specific yield, gpm/ft}}$$

$$\text{Drawdown, ft} = \frac{215\text{ gpm}}{19.6\text{ gpm/ft}} = 10.97\text{ ft, round to }\mathbf{11.0\text{ ft}}$$

110. Calculate the well yield in gallons per minute given a drawdown of 5.5 ft and a specific yield of 57 gpm/ft.

$$\textbf{Well yield, gpm} = \textbf{(Specific yield, gpm/ft)(Drawdown, ft)}$$

$$\text{Well yield, gpm} = (57\text{ gpm/ft})(5.5\text{ ft}) = 313.5\text{ gpm, round to }\mathbf{310\text{ gpm}}$$

111. The pumping water level in a well is 55 ft. If the static water level (when pump is off) is 38 ft, what is the drawdown for the well?

The equation is:

$$\textbf{Drawdown, ft} = \textbf{Pumping water level, ft} - \textbf{Static water level, ft}$$

$$\text{Drawdown, ft} = 55 \text{ ft} - 38 \text{ ft} = \textbf{17 ft}$$

112. Find the drawdown of a well if the well yields 408 gpm and the specific yield is 22.1.

Write the equation, arranging it to solve for the unknown, drawdown.

$$\textbf{Drawdown, ft} = \frac{\textbf{Well yield, gpm}}{\textbf{Specific yield, gpm/ft}}$$

$$\text{Drawdown, ft} = \frac{408 \text{ gpm}}{22.1 \text{ gpm/ft}} = 18.46 \text{ ft, round to } \textbf{18.5 ft}$$

113. The drawdown worksheet got wet and the operators were unable to read what the pumping water level was. If the static level in the well was 48.35 ft and the drawdown was 16.58 ft, what was the pumping water level in the well?

The equation is:

$$\textbf{Drawdown, ft} = \textbf{Pumping water level, ft} - \textbf{Static water level, ft}$$

Rearrange the equation to solve for pumping water level.

$$\text{Pumping water level, ft} = \text{Drawdown, ft} + \text{Static water level}$$

Substitute known values.

$$\text{Pumping water level, ft} = 16.58 \text{ ft} + 48.35 \text{ ft} = \textbf{64.93 ft}$$

114. The drawdown worksheet got wet again. The pumping water level in the well is 106.91 ft. If the drawdown was 14.58 ft, what was the static water level in the well?

The equation is:

Drawdown, ft = Pumping water level, ft – Static water level, ft

Rearrange the equation to solve for static water level.

Static water level, ft = Pumping water level, ft – Drawdown, ft

Substitute known values.

Static water level, ft = 106.91 ft – 14.58 ft = **92.33 ft**

HARDNESS PROBLEMS

115. A water sample contains 74 mg/L of calcium and 12 mg/L of magnesium. What is the total hardness as $CaCO_3$?

The equation is:

Total hardness, mg/L as $CaCO_3$ = Ca hardness, as $CaCO_3$ + Mg hardness, mg/L as $CaCO_3$

Total hardness, mg/L as $CaCO_3$ = 74 mg/L Ca + 12 mg/L Mg = **86 mg/L as $CaCO_3$**

116. A water sample contains 99 mg/L of calcium and 18 mg/L of magnesium. What is the total hardness as $CaCO_3$?

The equation is:

Total hardness, mg/L as $CaCO_3$ = Ca hardness, mg/L as $CaCO_3$ + Mg hardness, mg/L as $CaCO_3$

Total hardness, mg/L as $CaCO_3$ = 99 mg/L Ca + 18 mg/L Mg

= **117 mg/L as $CaCO_3$**

SOLUTION, PERCENT SOLUTION, AND MIXTURE PROBLEMS

117. How many gallons of a 7.5% solution must be mixed with a 2.2% solution to make exactly 400 gal of a 5.0% solution?

Solve the problem using the dilution triangle.

7.5%		2.8*	2.8 parts of the 7.5% solution is required for every 5.3 parts
	5.0%		
2.2%		2.5†	2.5 parts of the 2.2% solution is required for every 5.3 parts
		5.3 total parts	

$$\frac{2.8 \text{ parts } (400 \text{ gal})}{5.3 \text{ parts}} = 211 \text{ gal, round to } \mathbf{210 \text{ gal of the } 7.5\% \text{ solution}}$$

$$\frac{2.5 \text{ parts } (400 \text{ gal})}{5.3 \text{ parts}} = \frac{189 \text{ gal, round to } \mathbf{190 \text{ gal of the } 2.2\% \text{ solution}}}{400 \text{ gal (added as a cross check of math)}}$$

To make the 400 gal of the 5.0% solution, mix 210 gal of the 7.5% solution with 190 gal of the 2.2% solution.

* This number is calculated by subtracting diagonally 2.2% from 5.0%, absolute (negative sign is dropped).
† This number is calculated by subtracting diagonally 7.5% from 5.0%.

118. How many pounds of high test hypochlorite (HTH) are needed to make exactly 200 gal of a 7.0% HTH solution?

The equation is:

$$\textbf{\% HTH solution} = \frac{\textbf{(lb HTH)(100\%)}}{\textbf{(Number of gal)(8.34 lb/gal)}}$$

Rearranging the equation:

$$\text{Number of lb HTH} = (\%\text{ solution})(\text{Number of gal})(8.34 \text{ lb/gal})/100\%$$

$$\text{Number of lb HTH} = (7.0\%\text{ solution})(200 \text{ gal})(8.34 \text{ lb/gal})/100\%$$

$$\text{Number of lb HTH} = 117 \text{ lb of HTH, round to } \mathbf{120 \text{ lb of HTH}}$$

WATER CONSUMPTION

119. A water treatment plant has to shut down for some emergency repairs. Given the following data, will the water plant have enough water in storage to supply the community while making their repairs?

Storage tank level is 26.75 ft and is 120 ft in diameter

Pumping rate, out = 1,375 gpm

Time for repair is 24 hr maximum

First, determine the amount of water in gallons.

$$\text{Volume, gal} = (0.785)(\text{Diameter})^2(\text{Height})(7.48 \text{ gal/ft}^3)$$

$$\text{Volume, gal} = (0.785)(120 \text{ ft})(120 \text{ ft})(26.75 \text{ ft})(7.48 \text{ gal/ft}^3)$$

$$\text{Volume, gal} = 2{,}261{,}817 \text{ gal}$$

Next, determine the water supply in hours.

The equation is:

$$\textbf{Supply, hr} = \frac{\textbf{Storage volume, gal}}{\textbf{(Pumping rate, gpm)(60 min/hr)}}$$

$$\text{Supply, hr} = \frac{2{,}261{,}817 \text{ gal}}{(1{,}375 \text{ gpm})(60 \text{ min/hr})} = 27.4 \text{ hr}$$

Water plant's time = Supply hours – Repair hours = 27.4 hr – 24 hr = **+3.4 hr**

Because supply time is larger than repair time, they have enough time for their repairs.

120. A water plant serves 1,250,000 people. If it treats an average of 150 mgd, what are the gallons per capita per day? Note: A capita = 1 person.

First, convert 150 mgd to million gallons.

$$\text{Number of gal} = (150\ \text{mgd})(1{,}000{,}000\ \text{gal/mil gal}) = 150{,}000{,}000\ \text{gal}$$

$$\textbf{Gallons per capita per day (GPCPD)} = \frac{\textbf{Volume, gal/d}}{\textbf{Population served/d}}$$

$$\text{GPCPD} = \frac{150{,}000{,}000\ \text{gal/d}}{1{,}250{,}000\ \text{capita/d}} = \textbf{120 GPCD, or 120 gal per person per day}$$

Water Treatment Grade 1 Test

(answers on p. 54)

1. Find the detention time for a treatment plant given the following information:
   - Five flocculation basins each 56 ft by 18 ft with a water depth of 11.5 ft
   - A sedimentation basin that is 325 ft long, 68 ft wide, and has an average water depth of 11 ft
   - Ten filters each 40 ft by 30 ft (each measured to nearest foot) with an average water depth of 12 ft
   - Flow of 17.4 mgd

2. The iron content of a raw water is 1.32 mg/L. What is the percent removal if the finished water contains 0.08 mg/L iron?

3. How many gallons of a 36% solution must be mixed with a 19% solution to make exactly 500 gal of a 24% solution?

4. A 61.5% calcium hypochlorite solution is used to treat 5.8 mgd. If 237 lb/d of calcium hypochlorite is used, what is the chlorine dose in milligrams per liter?

5. Find the drawdown of a well if the well yields 278 gpm and the specific yield is 21.1.

6. A 2.0-ft diameter pipe that is 1,800 ft long was disinfected with chlorine. If 17.9 lb of chlorine were used, what was the dosage in milligrams per liter?

7. Thirty pounds of lime are mixed in a 100-gal drum that contains 75 gal of water. What is the percent of lime in the slurry?

8. A soda ash tank is conical at the bottom and cylindrical at the top. If the diameter of the cylinder is 15 ft with a depth of 30 ft and the cone depth is 12 ft, what is the volume of the tank in cubic feet?

9. A treatment plant processes an average of 4,125 gpm. If the lime dosage is 101 g/min, how many pounds of lime will the plant use in 1 month (30 days)? What is the dosage in milligrams per liter?

10. A filter has a surface area of 860 ft. What is the filtration rate in gallons per minute per square foot if it receives a flow of 5,200 gpm?

11. A polymer solution weighs 14.67 g per 10.0 mL. The water treatment plant is using 29.0 mL/min of the polymer for treating 16.5 ft^3/s. What is the polymer dosage?

12. If 95 lb of HTH (60.0% available chlorine) are mixed with 100 gal of water, what percent solution of 100% available chlorine is this? What percent is the calcium hypochlorite solution? Assume solution weighs 8.34 lb/gal.

Solutions to Water Treatment Grade 1 Test

1. First, determine the number of gallons in the five flocculation basins and the sed basin. The equation is:

$$\textbf{Volume, gal} = \textbf{(Length)(Width)(Depth)(7.48 gal/ft}^3\textbf{)(5 basins)}$$

$$\text{Volume, gal, floc basins} = (56\text{ ft})(19.5\text{ ft})(11.5\text{ ft})(7.48\text{ gal/ft}^3)(5\text{ basins}) = 469{,}669\text{ gal}$$

$$\text{Volume, gal in sed basin} = (325\text{ ft})(68\text{ ft})(11\text{ ft})(7.48\text{ gal/ft}^3) = 1{,}818{,}388\text{ gal}$$

$$\text{Volume, gal in filters} = (40\text{ ft})(30\text{ ft})(12)(7.48\text{ gal/ft}^3)(10\text{ filters}) = \frac{1{,}077{,}120\text{ gal}}{3{,}365{,}177\text{ gal}}$$

Next, convert million gallons per day to gallons per hour.

$$\textbf{(17.4 mgd)(1 d/24 hr)(1,000,000/1 M)} = \textbf{725,000 gph}$$

Write the equation with units asked for in the question.

$$\textbf{Detention time, hr} = \frac{\textbf{Volume, gal}}{\textbf{Flow rate, gph}}$$

$$\text{Detention time, hr} = \frac{3{,}365{,}177\text{ gal}}{725{,}000\text{ gph}} = \textbf{4.6 hr}$$

2. The equation is:

$$\textbf{Percent iron removal} = \frac{\textbf{In} - \textbf{Out}}{\textbf{In}} \times \textbf{100\%}$$

$$\text{Percent iron removal} = \frac{1.32\text{ mg/L} - 0.08}{1.32\text{ mg/L}} \times 100\% = \textbf{94\% removal efficiency}$$

3. Solve the problem using the dilution triangle.

36%	5	5 parts of the 36% solution are required for every 17 parts
24%		
19%	12	12 parts of the 19% solution are required for every 17 parts
	17 total parts	

$$\frac{5 \text{ parts (500 gal)}}{17 \text{ parts}} = \textbf{147 gal of the 36\% solution}$$

$$\frac{12 \text{ parts (500 gal)}}{17 \text{ parts}} = \frac{\textbf{353 gal of the 19\% solution}}{500 \text{ gal}}$$

To make the 500 gal of the 24% solution, mix 147 gal of the 36% solution with 353 gal of the 19% solution.

4. First, determine the number of pounds of chlorine available.

Available chlorine = (lb used)(% purity) = (237 lb/d)(61.5%/100%) = 145.755 lb/d

The equation is:

Pounds per day = (mgd)(Dosage, mg/L)(8.34 lb/gal)

Rearrange and solve for dosage.

$$\text{Chlorine dosage, mg/L} = \frac{145.755 \text{ lb/d}}{(5.8 \text{ mgd})(8.34 \text{ lb/d})} = \textbf{3.0 mg/L, calcium hypochlorite}$$

5. Write the equation, arranging it to solve for the unknown, drawdown.

$$\textbf{Drawdown, ft} = \frac{\textbf{Well yield, gpm}}{\textbf{Specific yield, gpm/ft}}$$

$$\text{Drawdown, ft} = \frac{278 \text{ gpm}}{21.1 \text{ gpm/ft}} = \textbf{13.2 ft}$$

6. First, find the volume in cubic feet for the pipe.

The equation is:

$$\textbf{Volume, ft}^3 = \textbf{(0.785)(Diameter)}^2\textbf{(Length, ft)}$$

$$\text{Volume, ft}^3 = (0.785)(2.0 \text{ ft})(2.0 \text{ ft})(1{,}800 \text{ ft}) = 5{,}652 \text{ ft}$$

Then, determine the number of gallons.

$$\text{Number of gal} = (5{,}652 \text{ ft})(7.48 \text{ gal/ft}^3) = 42{,}277 \text{ gal}$$

Convert the number of gallons to million gallons.

$$\text{Million gallons} = \frac{42{,}277 \text{ gal}}{1{,}000{,}000/1 \text{ M}} = 0.042277 \text{ mil gal}$$

Now, calculate the dosage.

$$\textbf{Chlorine dosage, mg/L} = \frac{\textbf{lb chlorine}}{\textbf{(mil gal)(8.34 lb/gal)}}$$

$$\text{Chlorine dosage, mg/L} = \frac{17.9 \text{ lb}}{(0.042277 \text{ mil gal})(8.34 \text{ lb/gal})}$$

$$= \textbf{50.767 mg/L, round to 51 mg/L Cl}_2$$

7. $$\text{Percent lime} = \frac{(30 \text{ lb})(100\%)}{30 \text{ lb} + (8.34 \text{ lb/gal})(75 \text{ gal})} = \frac{(30 \text{ lb})(100\%)}{30 \text{ lb} + 625.5 \text{ lb}} = \frac{(30 \text{ lb})(100\%)}{655.5 \text{ lb}}$$

$$= \textbf{4.58 lime slurry, round to 5\%}$$

8. First, find the volume of the cone in cubic feet.

$$\textbf{Volume, ft}^3 = \textbf{1/3(0.785)(Diameter)}^2\textbf{(Depth)}$$

$$\text{Volume, ft}^3 = 1/3(0.785)(15 \text{ ft})(15 \text{ ft})(12 \text{ ft}) = 706 \text{ ft}^3$$

Next, find the volume of the cylindrical part of the tank.

$$\text{Volume} = (0.785)(\text{Diameter})^2(\text{Depth})$$

$$\text{Volume} = (0.785)(15 \text{ ft})(15 \text{ ft})(30 \text{ ft}) = 5{,}299 \text{ ft}^3$$

Last, add the two volumes for the answer.

$$\text{Total volume, ft}^3 = 706 \text{ ft}^3 + 5{,}299 \text{ ft}^3 = \textbf{6,000 ft}^3$$

9. First, convert gallons per minute to million gallons per day.

$$(4{,}125 \text{ gpm})(1{,}440 \text{ min/d})(1 \text{ M}/1{,}000{,}000) = 5.94 \text{ mgd}$$

Next, convert grams per minute of lime to pounds per day.

$$(101 \text{ g/min})(1 \text{ lb}/454 \text{ g})(1{,}440 \text{ min/d}) = 320.35 \text{ lb/d}$$

Find number of pounds in 30 days.

$$(320.35 \text{ lb/d})(30 \text{ d/month}) = \mathbf{9{,}610 \text{ lb/month of lime}}$$

To find the dosage, use the "pounds" equation and rearrange to solve for dosage.

The equation is:

$$\textbf{Pounds per day} = \textbf{(mgd)(Dosage, mg/L)(8.34 lb/gal)}$$

Last, determine the dosage.

$$\textbf{Dosage, mg/L} = \frac{\textbf{lb/d lime}}{\textbf{(mgd)(8.34 lb/gal)}}$$

$$\text{Lime dosage, mg/L} = \frac{320.35 \text{ lb/d}}{(5.94 \text{ mgd})(8.34 \text{ lb/gal})} = \mathbf{6.47 \text{ mg/L of lime}}$$

10. The equation is:

$$\textbf{Filtration rate} = \frac{\textbf{Flow rate, gpm}}{\textbf{Filter surface area, ft}^2}$$

$$\text{Filtration rate} = \frac{5{,}200 \text{ gpm}}{860 \text{ ft}^2} = \mathbf{6.0 \text{ gpm/ft}^2}$$

11. First, determine the number of grams per milliliter.

$$14.67 \text{ g}/10 \text{ mL} = 1.467 \text{ g/mL}$$

Then, find number of pounds per gallon for the polymer.

$$(1.467 \text{ g/mL})(3{,}785 \text{ mL/gal})(1 \text{ lb}/454 \text{ g}) = 12.23 \text{ lb/gal}$$

Next, convert cubic feet per second to million gallons per day.

$$(16.5 \text{ ft}^3\text{/s})(86{,}400 \text{ s/d})(7.48 \text{ gal/ ft}^3)(1 \text{ M}/1{,}000{,}000) = 10.66 \text{ mgd}$$

(continued)

Now, find the number of pounds per day of polymer usage.

$$(29.0 \text{ mL/min})(1{,}440 \text{ min/d})(1 \text{ gal}/3{,}785 \text{ mL})(12.23 \text{ lb/gal}) = 134.93 \text{ lb/d}$$

Next, determine the dosage.

$$\textbf{Dosage, mg/L} = \frac{\textbf{lb/d}}{\textbf{(mgd)(8.34 lb/gal)}}$$

$$\text{Dosage, mg/L} = \frac{134.93 \text{ lb/d}}{(10.66 \text{ mgd})(8.34 \text{ lb/gal})} = \textbf{1.52 mg/L of polymer}$$

12. The equation is:

$$\textbf{Percent HTH solution} = \frac{\textbf{(lb HTH)(100\%)}}{\textbf{(Number of gal)(8.34 lb/gal)}}$$

$$\text{Percent HTH solution} = \frac{(95 \text{ lb HTH})(100\%)}{(100 \text{ gal})(8.34 \text{ lb/gal})}$$

$$= 11.39\%, \text{ round to } \textbf{10\% HTH solution}$$

The question asks for two results, so rounding off two times is required, but do not use the rounded off number in the first answer to get the second answer.

Percent available chlorine solution = (11.39%)(60.0%/100%)

= **6.83%, round to 7% available chlorine**

Note: The limiting number in this problem is 100 gal. It has only one significant figure.

CHAPTER 2

WATER TREATMENT

Grade 2

Students preparing for the grade 3 and grade 4 water treatment certification tests should also understand these problems.

CONVERSION PROBLEMS

1. Convert 25 gal to cubic feet.

The equation is:

$$\mathbf{(Number\ of\ gal)\frac{(1\ ft^3)}{7.48\ gal} = Number\ of\ ft^3}$$

$$(25\text{ gal})\frac{(1\text{ ft}^3)}{7.48\text{ gal}} = 3.34\text{ ft}^3\text{, round to }\mathbf{3.3\ ft^3}$$

2. How much does 354 gal of water weigh?

$$(354\text{ gal})\frac{(8.34\text{ lb})}{\text{gal}} = 2{,}952.36\text{ lb, round to }\mathbf{2{,}950\ lb}$$

3. Convert 12.5 ft³/s to million gallons per day.

$$\frac{(12.5\ \text{ft}^3)\ (60\ \text{s})\ (1{,}440\ \text{min})\ (7.48\ \text{gal})\ (1\ \text{mil gal})}{\text{sec}\quad \text{min}\quad \text{day}\quad \text{ft}^3\quad 1{,}000{,}000\ \text{gal}} = \mathbf{8.08\ mgd}$$

4. Convert 1.84 mgd to cubic feet per second.

$$\frac{(1.84\ \text{mgd})\ (1{,}000{,}000\ \text{gal})\ (1\ \text{ft}^3)\quad (1\ \text{d})\quad (1\ \text{min})}{1\ \text{mil gal}\quad 7.48\ \text{gal}\quad 1{,}440\ \text{min}\quad 60\ \text{s}} = \mathbf{2.85\ ft^3/s}$$

5. Convert 7,020,451 ft³ to acre-feet.

$$\frac{7{,}020{,}451\ \text{ft}^3}{43{,}560\ \text{ft}^3/\text{acre-ft}} = \mathbf{161.1674\ acre\text{-}ft}$$

6. Convert 68,428 gal to cubic feet.

$$(68{,}428\ \text{gal})\ \frac{(1\ \text{ft}^3)}{7.48\ \text{gal}} = 9{,}148.13\ \text{ft}^3\text{, round to } \mathbf{9{,}150\ ft^3}$$

7. Convert 2,286 gpm to cubic feet per second.

$$\frac{2{,}286\ \text{gpm}}{(60\ \text{s/min})(7.48\ \text{gal/ft}^3)} = \mathbf{5.09\ ft^3/s}$$

8. How many gallons are there in 8,492 ft³?

$$\frac{(7.48\text{ gal})}{1\text{ ft}^3}(8{,}492\text{ ft}^3) = 63{,}520.16\text{ gal, round to }\mathbf{63{,}500\text{ gal}}$$

9. Convert 2.4 mgd to cubic feet per second.

The equation is:

$$\frac{(\text{Number of mgd})\ (1{,}000{,}000\text{ gal})\ (1\text{ ft}^3)\quad (1\text{ d})\quad (1\text{ min})}{1\text{ mil gal}\quad 7.48\text{ gal}\quad 1{,}440\text{ min}\quad 60\text{ s}} = \text{Number of ft}^3\text{/s}$$

$$\frac{(2.4\text{ mgd})\ (1{,}000{,}000\text{ gal})\ (1\text{ ft}^3)\quad (1\text{ d})\quad (1\text{ min})}{1\text{ mil gal}\quad 7.48\text{ gal}\quad 1{,}440\text{ min}\quad 60\text{ s}} = \mathbf{3.7\text{ ft}^3\text{/s}}$$

10. Convert 15 ft³/s to million gallons per day.

The equation is:

$$\frac{(\text{Number ft}^3)\ (60\text{ s})\ (1{,}440\text{ min})\ (7.48\text{ gal})\ (1\text{ mil gal})}{\text{sec}\quad \text{min}\quad \text{day}\quad \text{ft}^3\quad 1{,}000{,}000\text{ gal}} = \text{Number of mgd}$$

$$\frac{(15\text{ ft}^3)\ (60\text{ s})\ (1{,}440\text{ min})\ (7.48\text{ gal})\ (1\text{ mil gal})}{\text{sec}\quad \text{min}\quad \text{day}\quad \text{ft}^3\quad 1{,}000{,}000\text{ gal}} = \mathbf{9.7\text{ mgd}}$$

11. How many million gallons are there in 145 acre-ft?

$(145\text{ acre-ft})(43{,}560\text{ ft}^3/\text{acre-ft})(7.48\text{ gal/ft}^3)(1\text{ M}/1{,}000{,}000) =$ **47.2 mil gal**

12. How many gallons are there in 15 acre-ft?

First, convert acre feet into cubic feet.

$$(15 \text{ acre-ft})(43{,}560 \text{ ft}^3/\text{acre-ft}) = 653{,}400 \text{ ft}^3$$

$$(653{,}400 \text{ ft}^3)(7.48 \text{ gal/ft}^3) = 4{,}887{,}432 \text{ gal, round to } \mathbf{4{,}900{,}000 \text{ gal}}$$

13. Convert 3,425 gpm to cubic feet per second.

The equation is:

$$\frac{\textbf{Number of ft}^3\textbf{/s}}{\textbf{(60 s/min)(7.48 gal/ft}^3\textbf{)}} = \textbf{Number gpm}$$

$$\text{Number of ft}^3\text{/s} = \frac{3{,}425 \text{ gpm}}{(60 \text{ s/min})(7.48 \text{ gal/ft}^3)} = 7.63 \text{ ft}^3\text{/s}$$

TEMPERATURE CONVERSION PROBLEMS

14. Convert 30°F to degrees Celsius.

The equation for Celsius is:

$$\mathbf{°C = {}^5\!/_9(°F - 32)}$$

$$°C = {}^5\!/_9(30°F - 32) = {}^5\!/_9(-2) = \mathbf{-1°C}$$

15. Convert 200°C to degrees Fahrenheit.

The equation for Fahrenheit is:

$$\mathbf{°F = ({}^9\!/_5)°C + 32}$$

$$°F = ({}^9\!/_5)200 + 32 = 392, \text{ round to } \mathbf{400°F}$$

16. Convert 48°C to degrees Fahrenheit.

$$°F = (9/5)48 + 32 = 118.4, \text{ round to } \mathbf{120°F}$$

PERCENT AND PARTS PER MILLION CONVERSION PROBLEMS

17. Convert a solution that has 305,000 ppm to percent.

Note that a 1% solution = 10,000 ppm

$$\frac{305{,}000 \text{ ppm}}{10{,}000 \text{ ppm}/1\%} = 30.5\% \text{ solution}$$

18. A solution was found to be 0.48% alum. How many parts per million of alum are in the solution?

If a 1% solution has 10,000 ppm, a 0.48% solution will have:

$$(0.48\%)(10{,}000 \text{ ppm}/1\%) = \mathbf{4{,}800 \text{ ppm alum}}$$

19. Convert a solution that has 259,000 ppm to percent.

Note that a 1% solution = 10,000 ppm

$$\frac{259{,}000 \text{ ppm}}{10{,}000 \text{ ppm}/1\%} = 25.9\% \text{ solution}$$

PERCENT CALCULATIONS

20. What is the percent removal across a settling basin if the influent is 17.60 ntu and the effluent is 1.29 ntu?

The equation is:

$$\textbf{Percent ntu removal} = \frac{\textbf{In} - \textbf{Out}}{\textbf{In}} \times \textbf{100\%}$$

$$\text{Percent ntu removal} = \frac{17.60 \text{ ntu} - 1.29 \text{ ntu}}{17.60 \text{ ntu}} \times 100\% = \frac{16.31 \text{ ntu}}{17.60 \text{ ntu}} \times 100\%$$

$$\text{Percent ntu removal} = 0.9267 \times 100 = \textbf{92.7\% removal of ntu}$$

21. The iron content of a raw water is 1.13 mg/L. What is the percent removal if the finished water contains 0.09 mg/L iron?

The equation is:

$$\textbf{Percent iron removal} = \frac{\textbf{In} - \textbf{Out}}{\textbf{In}} \times \textbf{100\%}$$

$$\text{Percent iron removal} = \frac{1.13 \text{ mg/L} - 0.09 \text{ mg/L}}{1.13 \text{ mg/L}} \times 100\%$$

$$= 92.0\%, \text{ round to } \textbf{90\% removal efficiency}$$

22. What is the percent removal across a settling basin if the influent is 41.6 ntu and the effluent is 2.8 ntu?

The equation is:

$$\textbf{Percent ntu removal} = \frac{\textbf{Influent ntu} - \textbf{Effluent ntu}}{\textbf{Influent ntu}} \times \textbf{100\%, or } \frac{\text{In} - \text{Out}}{\text{In}} \times 100\%$$

$$\text{Percent ntu removal} = \frac{41.6 \text{ ntu} - 2.8 \text{ ntu}}{41.6 \text{ ntu}} \times 100\% = \frac{38.8 \text{ ntu}}{41.6 \text{ ntu}} \times 100\%$$

$$\text{Percent ntu removal} = 0.93 \times 100\% = \textbf{93\% removal of ntu}$$

23. Ten pounds of lime are mixed in a 55-gal drum that contains 50 gal of water. What is the percent of lime in the slurry?

$$\text{Percent of lime} = \frac{(10 \text{ lb})(100\%)}{10 \text{ lb} + (8.34 \text{ lb/gal})(50 \text{ gal})} = \frac{(10 \text{ lb})(100\%)}{10 \text{ lb} + 417 \text{ lb}} = \frac{(10 \text{ lb})(100\%)}{427 \text{ lb}}$$

$$= \textbf{2\% lime slurry}$$

24. The iron content of a raw water is 2.28 mg/L. What is the percent removal if the finished water contains 0.13 mg/L iron?

The equation is:

$$\textbf{Percent iron removal} = \frac{\textbf{In} - \textbf{Out}}{\textbf{In}} \times \textbf{100\%}$$

$$\text{Percent iron removal} = \frac{2.28 \text{ mg/L} - 0.13}{2.28 \text{ mg/L}} \times 100\% = \textbf{94\% removal efficiency}$$

AVERAGE CALCULATIONS

25. What is the average number of pounds of lime used per day given the following data? Assume three significant figures.

Mon.	Tues.	Wed.	Thurs.	Fri.	Sat.	Sun.
254	241	261	250	236	240	262

The equation is:

$$\textbf{Average lime used, lb/d} = \frac{\textbf{Sum of lime used each day, lb}}{\textbf{Total time, d}}$$

Substituting:

$$\text{Average lime used, lb/d} = \frac{254 + 241 + 261 + 250 + 236 + 240 + 262}{7\text{ d}} = \textbf{249 lb/d lime}$$

26. What is the average million gallons per day production for a treatment plant given the following data?

Mon.	Tues.	Wed.	Thurs.	Fri.	Sat.	Sun.
3.8	3.5	3.1	3.0	3.2	3.2	3.5

The equation is:

$$\textbf{Average mgd produced} = \frac{\textbf{Sum of mgd used each day}}{\textbf{Total time, d}}$$

$$\text{Average mgd produced} = \frac{3.8 + 3.5 + 3.1 + 3.0 + 3.2 + 3.2 + 3.5}{7\text{ d}}$$

$$= 3.33\text{ mgd, round to } \textbf{3.3 mgd}$$

AREA PROBLEMS

27. What is the area of a drying bed that is 250 ft long and 45 ft wide?

The equation is:

$$\textbf{Area} = \textbf{(Length)(Width)}$$

$$\text{Area} = (250\text{ ft})(45\text{ ft}) = 11{,}250\text{ ft}^2\text{, round to }\mathbf{11{,}000\text{ ft}^2}$$

28. What area does a tank occupy if it has a diameter of 48.0 ft?

The equation is:

$$\textbf{Area} = \pi r^2$$

or use:

$$\textbf{Area} = \textbf{(0.785)(Diameter)}^2$$

Both equations are used throughout this book.

$$\text{Area of tank} = (0.785)(48.0\text{ ft})(48.0\text{ ft}) = 1{,}809\text{ ft}^2\text{, round to }\mathbf{1{,}810\text{ ft}^2}$$

29. If the surface area of a rectangular drying bed is 22,500 ft² and one side is 50 ft, what is the length of the other side? Assume 50 ft is measured to the nearest foot.

First, write the equation for finding the area of a rectangle.

$$\textbf{Area} = \textbf{(Length)(Width)}$$

Since the area is 22,500 ft and one side is 50 ft, let's assume that 50 ft is the width (remember, by convention, the width is the shorter of the two measurements). Then, solve for the length by rearranging the equation as follows:

$$\text{Length, ft} = \frac{\text{Area of }22{,}500\text{ ft}^2}{(50\text{ ft})} = \mathbf{450\text{ ft}}$$

30. What is the diameter of a tank if the surface area is 4,542 ft²?

The equation for finding the area of a circular tank is:

$$\textbf{Area} = \textbf{(0.785)(Diameter)}^2 \text{ or } \textbf{Area} = (\pi)(\mathbf{r})^2 = \textbf{(3.14)(r)}^2$$

Solve for the diameter by rearranging the equation as shown.

$$\text{Diameter}^2 = \frac{\text{Area}}{0.785} = \frac{4{,}542 \text{ ft}^2}{0.785} = 5{,}786 \text{ ft}^2$$

Then, take the square root of 5,786 ft^2.

Diameter = **76.1 ft**

31. What is the surface area of a pond that is 302 ft long and 179 ft wide?

Area, ft^2 = (Length)(Width) = (302 ft)(179 ft) = 54,058 ft^2, round to **54,100 ft²**

CIRCUMFERENCE PROBLEMS

32. If the circumference of a tank is 210 ft, what is the radius?

Circumference = π(Diameter)

210 ft = (3.14)(Diameter)

Diameter = 210 ft/3.14 = 66.9 ft

The radius is equal to the diameter/2, or

Radius = 66.9/2 = 33.45 ft, round to **33 ft**

33. What is the circumference, in feet, of a tank that is 48 ft in diameter?

The equation is:

$$\textbf{Circumference} = \pi\textbf{(Diameter)}$$

$$\text{Circumference} = (3.14)(48 \text{ ft}) = 150.72 \text{ ft, round to } \textbf{150 ft}$$

VOLUME PROBLEMS

34. A circular tank has a radius of 24 ft and is 30.0 ft high. What is the capacity of the tank in cubic feet?

The volume equation for a circular tank is:

$$\textbf{Volume} = \pi r^2\textbf{(Height)}$$

where r is the radius.

$$\text{Volume of tank in ft}^3 = (3.14)(24 \text{ ft})(24 \text{ ft})(30.0 \text{ ft}) = 54{,}259.2 \text{ ft}^3\text{, round to } \textbf{54,000 ft}^3$$

35. What is the volume in cubic feet of two basins with the following dimensions: 200 ft by 30 ft by 6 ft and 180 ft by 26 ft by 6 ft?

The equation is:

$$\textbf{Volume} = \textbf{(Length)(Width)(Depth)}$$

Solve for each basin separately, then add the results.

$$\text{Basin 1} = (200 \text{ ft})(30 \text{ ft})(6 \text{ ft}) = 36{,}000 \text{ ft}^3$$

$$\text{Basin 2} = (180 \text{ ft})(26 \text{ ft})(6 \text{ ft}) = 28{,}080 \text{ ft}^3$$

$$36{,}000 \text{ ft}^3 + 28{,}080 \text{ ft}^3 = 64{,}080 \text{ ft}^3\text{, round to } \textbf{64,000 ft}^3$$

36. A soda ash tank is conical at the bottom and cylindrical at the top. If the diameter of the cylinder is 18 ft with a depth of 32 ft and the cone depth is 15 ft, what is the volume of the tank in cubic feet?

First, find the volume of the cone in cubic feet.

The equation is:

$$\textbf{Volume, ft}^3 = \textbf{1/3(0.785)(Diameter)}^2\textbf{(Depth)}$$

$$\text{Volume, ft}^3 = 1/3(0.785)(18\text{ ft})(18\text{ ft})(15\text{ ft}) = 1{,}272\text{ ft}^3$$

Next, find the volume of the cylindrical part of the tank.

$$\text{Volume} = (0.785)(\text{Diameter})^2(\text{Depth}) = (0.785)(18\text{ ft})(18\text{ ft})(32\text{ ft}) = 8{,}139\text{ ft}^3$$

Last, add the two volumes.

$$\text{Total volume} = 1{,}272\text{ ft}^3 + 8{,}139\text{ ft}^3 = 9{,}411\text{ ft}^3\text{, round to } \mathbf{9{,}400\text{ ft}^3}$$

37. Determine the volume in cubic feet for a pipe with a diameter of 3.0 ft and a length of 14.3 mi.

First, determine the number of feet in 14.3 mi.

$$(5{,}280\text{ ft/mi})(14.3\text{ mi}) = 75{,}504\text{ ft}$$

The equation is:

$$\textbf{Volume, ft}^3 = \textbf{(0.785)(Diameter)}^2\textbf{(Length)}$$

$$\text{Volume, ft}^3 = (0.785)(3.0\text{ ft})(3.0\text{ ft})(75{,}504\text{ ft}) = 533{,}436\text{ ft}^3\text{, round to } \mathbf{530{,}000\text{ ft}^3}$$

38. What is the volume of a conical tank in cubic feet that is 12 ft in diameter and 11 ft in height?

The equation is:

$$\textbf{Volume} = \mathbf{1/3\pi r^2(Height\ or\ Depth)}$$

First, find the radius.

$$\text{Radius} = \text{Diameter/2 or D/2}$$

$$\text{Radius} = 12\text{ ft/2} = 6.0\text{ ft}$$

Using the equation, substitute the known quantities.

$$\text{Volume, ft}^3 = 1/3(3.14)(6.0\text{ ft})(6.0\text{ ft})(11\text{ ft}) = 414.48\text{ ft}^3\text{, round to } \mathbf{410\ ft^3}$$

39. A lime tank is conical at the bottom and cylindrical at the top. If the cylinder is 14 ft in diameter, has a depth of 24 ft, and the cone depth is 12 ft, what is the volume of the tank in cubic feet?

First, find the volume of the cone in ft^3.

$$\mathbf{Volume,\ ft^3 = 1/3\pi r^2(Depth)}$$

where the radius = Diameter/2 = 14 ft/2 = 7 ft

$$\text{Volume, ft}^3 = 1/3(3.14)(7\text{ ft})(7\text{ ft})(12\text{ ft}) = 615\text{ ft}^3$$

Next, find the volume of the cylindrical part of the tank.

$$\text{Volume} = \pi r^2(\text{Depth}) = (3.14)(7\text{ ft})(7\text{ ft})(24\text{ ft}) = 3{,}693\text{ ft}^3$$

Last, add the two volumes.

$$\text{Total volume, ft}^3 = 615\text{ ft}^3 + 3{,}693\text{ ft}^3 = 4{,}308\text{ ft}^3\text{, round to } \mathbf{4{,}300\ ft^3}$$

40. Find the volume in gallons for two sedimentation (sed) basins that are 360 ft by 75 ft and six filters that are 35 ft by 24 ft. The depth of water in the sed basins is 10.5 ft and in the filters is 9.5 ft.

First, find the volume in gallons for the two sed basins.

$$\textbf{Volume, gal} = \textbf{(Length)(Width)(Depth)(7.48 gal/ft}^3\textbf{)(Number of sed basins)}$$

$$\text{Volume, gal} = (360\text{ ft})(75\text{ ft})(10.5\text{ ft})(7.48\text{ gal/ft}^3)(2\text{ sed basins}) = 4{,}241{,}160\text{ gal}$$

Next, find the volume in gallons for the filters, then add both volumes.

$$\text{Volume, gal} = (35\text{ ft})(24\text{ ft})(9.5\text{ ft})(7.48\text{ gal/ft}^3)(6\text{ filters}) = 358{,}142\text{ gal}$$

$$\text{Total volume, gal} = 4{,}241{,}160\text{ gal} + 358{,}142\text{ gal}$$

$$\text{Total volume, gal} = 4{,}599{,}302\text{ gal, round to } \mathbf{4{,}600{,}000\text{ gal}}$$

WATERSHED SUPPLY CALCULATIONS

41. Given the following data:

Average annual rainfall is 45 in.
Watershed is 325 mi^2
12% of the rainfall is collected

How many million gallons can the watershed supply each year and what would the average be per day?

First, find the number of inches collected from the rainfall.

$$(45\text{ in.})(12\%/100\%) = 5.4\text{ in.}$$

Convert the number of inches of rain collected to feet.

$$(5.4\text{ in.})(1\text{ ft}/12\text{ in.}) = 0.45\text{ ft}$$

Then, calculate the number of gallons in 1 acre that is covered by 0.45 ft of water.

$$(43{,}560\text{ ft}^2\text{/acre})(7.48\text{ gal/ft}^3)(0.45\text{ ft}) = 146{,}623\text{ gal/acre}$$

Convert this to million gallons.

$$(146{,}623\text{ gal/acre})(1\text{ M}/1{,}000{,}000) = 0.146623\text{ mil gal/acre}$$

Next, find the number of acres in 325 mi^2.

$$\text{Number of acres} = (640 \text{ acres/mi}^2)(325 \text{ mi}^2) = 208{,}000 \text{ acres}$$

Then, calculate the number of million gallons collected from the watershed.

(0.146623 mil gal/acre)(208,000 acres) = 30,498 mil gal/yr, round to **30,000 mil gal/yr**

Last, find the average daily supply the watershed can provide.

$$\frac{30{,}498 \text{ mil gal/yr}}{365 \text{ d/yr}} = 83.6 \text{ mgd, round to } \mathbf{84 \text{ mgd}}$$

42. A watershed that is 31 mi^2 receives an average of 18 in. of rain each year. Seven percent of the rain is collected for treatment. How many million gallons of water are available per year for the small community that the treatment plant supplies, and what is the average per capita consumption in gallons if 1,645 people use the treated water?

First, find the number of inches collected from the rainfall.

$$(18 \text{ in.})(7.0\%/100\%) = 1.26 \text{ in.}$$

Convert the number of inches of rain collected to feet.

$$(1.26 \text{ in.})(1 \text{ ft}/12 \text{ in.}) = 0.105 \text{ ft}$$

Then, calculate the number of gallons in 1 acre that is covered by 0.105 ft of water.

$$(43{,}560 \text{ ft}^2/\text{acre})(7.48 \text{ gal/ft}^3)(0.105 \text{ ft}) = 34{,}212 \text{ gal/acre}$$

Convert this to million gallons.

$$(34{,}212 \text{ gal/acre})(1 \text{ M}/1{,}000{,}000) = 0.034212 \text{ mil gal/acre}$$

Next, find the number of acres in 31 mi^2.

$$\text{Number of acres} = (640 \text{ acres/mi}^2)(31 \text{ mi}^2) = 19{,}840 \text{ acres}$$

(continued)

Then, calculate the number of million gallons collected from the watershed.

(0.034212 mil gal/acre)(19,840 acres) = 678.77 mil gal, round to **680 mil gal/yr**

Last, find the average per capita consumption in gallons for the community.

$$\frac{(678.77 \text{ mil gal/yr})(1{,}000{,}000/1 \text{ M})}{(1{,}645 \text{ people})(365 \text{ d/yr})} = 1{,}130, \text{ round to } \mathbf{1{,}100 \text{ gal/d/person}}$$

43. What was the rainfall for a watershed if the treatment facility collected 3,984 mil gal for the year, given the following data:

10% of the water was collected
Watershed is 84 mi^2

First, calculate the number of acres in 84 mi^2.

$$(640 \text{ acres/mi}^2)(84 \text{ mi}^2) = 53{,}760 \text{ acres}$$

Next, determine the number of gallons per acre.

$$\frac{(3{,}984 \text{ mil gal})(1{,}000{,}000/1 \text{ M})}{53{,}760 \text{ acres}} = 74{,}107 \text{ gal/acre}$$

Then, find the thickness this water would have over the acre, in inches.

$$\frac{(74{,}107 \text{ gal/acre})(12 \text{ in./ft})}{(43{,}560 \text{ ft}^2\text{/acre})(7.48 \text{ gal/ft}^3)} = 2.729 \text{ in.}$$ (***Note*:** This is only 10% of the rain.)

Last, calculate the amount of rainfall in inches.

(2.729 in.)(100%/10%) = 27.29, round to **30 in.**

DETENTION TIME CALCULATIONS

44. Calculate the detention time in hours for the portion of a treatment plant with the following characteristics: Five flocculation basins each 62.4 ft by 18.1 ft with an average water depth of 12.3 ft and a sedimentation (sed) basin that is 298.5 ft long, 83.2 ft wide, and has an average water depth of 10.5 ft; the flow is 20.9 mgd.

First, determine the number of gallons in the five flocculation basins and the sed basin.

The equation is:

$$\textbf{Volume, gal} = \textbf{(Length)(Width)(Depth)(7.48 gal/ft}^3\textbf{)(5 basins)}$$

Volume, in gallons, in floc basins =

$$(62.4 \text{ ft})(18.1 \text{ ft})(12.3 \text{ ft})(7.48 \text{ gal/ft}^3)(5 \text{ basins}) = 519{,}565 \text{ gal}$$

Volume, in gallons, in sed basin =

$$(298.5 \text{ ft})(83.2 \text{ ft})(10.5 \text{ ft})(7.48 \text{ gal/ft}^3) = \frac{1{,}950{,}557 \text{ gal}}{2{,}470{,}122 \text{ gal}}$$

Next, convert million gallons per day to gallons per hour.

$$(20.9 \text{ mgd})(1 \text{ d}/24 \text{ hr})(1{,}000{,}000/1 \text{ M}) = 870{,}833 \text{ gph}$$

Write the equation with units asked for in the question.

$$\textbf{Detention time, hr} = \frac{\textbf{Volume, gal}}{\textbf{Flow rate, gph}}$$

$$\text{Detention time, hr} = \frac{2{,}470{,}122 \text{ gal}}{870{,}833 \text{ gph}} = 2.836 \text{ hr, round to } \mathbf{2.8 \text{ hr}}$$

Detention times are theoretical because basins begin to fill with settled floc and other debris and because of the probability that there is some short-circuiting.

45. Calculate the detention time in hours for a water treatment plant with the following characteristics: Five flocculation basins each 48 ft by 20.3 ft with a water depth of 11.8 ft; a sedimentation basin that is 452 ft long, 79.9 ft wide, with an average water depth of 11 ft; and eight filters each 40.1 ft by 30.2 ft with an average water depth of 10.5 ft; the flow is 15.4 mgd.

First, determine the number of gallons in the five flocculation basins and the sed basin.

The equation is:

$$\textbf{Volume, gal} = \textbf{(Length)(Width)(Depth)(7.48 gal/ft}^3\textbf{)(5 basins)}$$

Volume, in gallons, in floc basins =

$$(48 \text{ ft})(20.3 \text{ ft})(11.8 \text{ ft})(7.48 \text{ gal/ft}^3)(5 \text{ basins}) = 430{,}022 \text{ gal}$$

Volume, in gallons, in sed basin =

$$(452 \text{ ft})(79.9 \text{ ft})(11 \text{ ft})(7.48 \text{ gal/ft}^3) = 2{,}971{,}526 \text{ gal}$$

Volume, in gallons, in filters =

$$(40.1 \text{ ft})(30.2 \text{ ft})(10.5 \text{ ft})(7.48 \text{ gal/ft}^3)(8 \text{ filters}) = \underline{760{,}908 \text{ gal}}$$

$$4{,}162{,}456 \text{ gal}$$

Next, convert million gallons per day to gallons per hour.

$$(15.4 \text{ mgd})(1 \text{ d}/24 \text{ hr})(1{,}000{,}000/1\text{M}) = 641{,}667 \text{ gph}$$

Write the equation with units asked for in the question.

$$\textbf{Detention time, hr} = \frac{\textbf{Volume, gal}}{\textbf{Flow rate, gph}}$$

$$\text{Detention time, hr} = \frac{4{,}162{,}456 \text{ gal}}{641{,}667 \text{ gph}} = 6.49 \text{ hr, round to } \mathbf{6.5\ hr}$$

PRESSURE PROBLEMS

46. What is the psi at the bottom of a clear well if the water level is 19.05 ft deep?

The equation is:

$$\textbf{psi} = \frac{\textbf{Depth, ft}}{\textbf{2.31 ft/psi}}$$

$$\text{psi} = \frac{19.05 \text{ ft}}{2.31 \text{ ft/psi}} = \textbf{8.25 psi}$$

47. A lake is 148 ft deep. What is the psi on the bottom?

An important number to commit to memory is 0.433 psi/ft, which is 1 psi/2.31 ft.

The equation would then be:

$$\textbf{psi} = \textbf{(Depth, ft)(0.433 psi/ft)}$$

$$\text{psi} = (148 \text{ ft})(0.433 \text{ psi/ft}) = \textbf{64.1 psi}$$

48. What is the pressure, in pounds per square foot, 17.34 ft below a lake's surface?

The equation is:

$$\textbf{Pressure} = \textbf{(Depth)(Density, 62.4 lb/ft}^3\textbf{)}$$

$$\text{Pressure} = (17.34 \text{ ft})(62.4 \text{ lb/ft}^3) = 1{,}082.0 \text{ lb/ft}^2\text{, round to } \textbf{1,080 lb/ft}^2$$

DENSITY CALCULATIONS

Density of a substance is the amount of mass for a given volume.

49. The density of an unknown substance is 4.25 g/cm³. How much space would this substance occupy in cubic millimeters if it weighed 9.45 lb?

First, convert the number of pounds to grams.

$$\textbf{Number of g} = \textbf{(Number of lb)(454 g/1 lb)}$$

Substituting:

$$\text{Number of g} = (9.45 \text{ lb})(454 \text{ g/1 lb}) = 4{,}290 \text{ g}$$

We know that 4.25 g of the substance occupies 1 cm^3 by knowing its density. To get the amount of space 4,290 g occupies, divide by the density.

$$\text{Space occupied by substance} = \frac{4{,}290 \text{ g}}{4.25 \text{ g/cm}^3} = 1{,}009 \text{ cm}^3\text{, round to } \mathbf{1{,}010\ cm^3}$$

50. What is the density in pounds per gallon of a 1.000-L solution that weighs 1.5 lb?

First, convert milliliters to gallons.

$$(1.000 \text{ L})(1 \text{ gal}/3.785 \text{ L}) = 0.2642$$

The equation is:

$$\textbf{Density} = \textbf{Mass/Volume}$$

$$\text{Density of solution} = 1.5 \text{ lb}/0.2642 \text{ gal} = 5.7 \text{ lb/gal}$$

51. Find the number of gallons per cubic foot of a solution if it weighs 84.8 lb/ft³.

Divide the weight of the substance by 8.34 lb/gal to compare the substance to water.

$$\text{Density} = \frac{84.8 \text{ lb/ft}^3}{8.34 \text{ lb/gal}} = \mathbf{10.2\ gal/ft^3}$$

52. A substance weighs 605 g and occupies a space of 265 cm³. What is its density in grams per cubic centimeter?

The equation is:

$$\textbf{Density} = \textbf{Mass/Volume}$$

$$\text{Density} = 605 \text{ g}/265 \text{ cm}^3 = \mathbf{2.28\ g/cm^3}$$

SPECIFIC GRAVITY OF LIQUIDS

Specific gravity compares the density of one substance to that of another. Water is the standard for liquids.

53. The density of an unknown liquid is 62.8 lb/ft³. What is the specific gravity?

Water has a density of 62.4 lb/ft³. Divide the density of the unknown by the density of water.

The equation is:

$$\textbf{Specific gravity} = \textbf{Density of substance/Density of water}$$

$$\text{Specific gravity of unknown substance} = \frac{62.8 \text{ lb/ft}^3}{62.4 \text{ lb/ft}^3} = 1.006, \text{ round to } \mathbf{1.01\ sp\ gr}$$

54. What is the specific gravity for a solution that weighs 9.84 lb/gal?

The density of water can also be expressed as lb/gal, or 8.34 lb/gal.

$$\text{Specific gravity} = \frac{9.84 \text{ lb/gal}}{8.34 \text{ lb/gal}} = \mathbf{1.18\ sp\ gr}$$

55. The specific gravity (sp gr) of a liquid solution is 1.12. How many pounds does 1 ft^3 weigh?

The equation is:

$$\textbf{lb/ft}^3 = \textbf{(sp gr)(62.4 lb/ft}^3\textbf{)}$$

lb/gal of liquid solution = (1.12)(62.4 lb/ft^3) = **69.9 lb/ft^3 for liquid solution**

FLOW PROBLEMS

56. What is the velocity of flow in feet per second for a 3.0-in. diameter pipe if it delivers 72 gpm?

First, convert the number of gallons per minute to cubic feet per second.

$$\text{Number of ft}^3\text{/s} = \frac{72 \text{ gpm}}{(7.48 \text{ gal/ft}^3)(60 \text{ s/min})} = 0.16 \text{ ft}^3\text{/s}$$

Next, convert the diameter from inches to feet.

$$\text{Number of ft} = (3.0 \text{ in.})(1 \text{ ft/12 in.}) = 0.25$$

The equation is:

$$\textbf{(Q)Flow, ft}^3\textbf{/s} = \textbf{(Area, ft}^2\textbf{)(Velocity, ft/s)}$$

where the area = (0.785)(Diameter)^2

$$0.16 \text{ ft}^3\text{/s} = (0.785)(0.25 \text{ ft})(0.25 \text{ ft})(\text{Velocity, ft/s})$$

Rearrange and solve for the flow in feet per second.

$$\text{Flow, ft/s} = \frac{0.16 \text{ ft}^3\text{/s}}{(0.785)(0.25 \text{ ft})(0.25 \text{ ft})} = 3.26 \text{ ft/s, round to } \textbf{3.3 ft/s}$$

57. Flow through a 5.2-ft wide channel is 35.5 ft^3/s. If the velocity is 1.7 ft/s, what is the depth of the water in the channel?

The equation is:

$$\textbf{Q (Flow)} = \textbf{(Area)(Velocity)}$$

$$35.5\ ft^3/s = (5.2\ ft)(x\ ft,\ depth)(1.7\ ft/s)$$

Now, solve for depth.

$$x\ ft,\ depth = \frac{35.5\ ft^3/s}{(5.2\ ft)(1.7\ ft/s)} = \textbf{4.0 ft in depth}$$

CALCULATIONS OF CHLORINE DOSE, DEMAND, AND RESIDUAL

58. If the chlorine residual is 1.72 mg/L and the chlorine demand is 1.30 mg/L, what is the chlorine dose?

The equation is:

Chlorine dose = Chlorine demand + Chlorine residual

Chlorine dose = 1.72 mg/L + 1.30 mg/L = **3.02 mg/L**

59. If the chlorine dose is 4.65 mg/L and the chlorine residual is 1.83 mg/L, what is the chlorine demand?

The equation is:

Chlorine dose = Chlorine demand + Chlorine residual

Rearrange and solve for chlorine demand.

Chlorine demand = Chlorine dose – Chlorine residual

Chlorine demand = 4.65 mg/L – 1.83 mg/L = **2.82 mg/L**

DOSAGE PROBLEMS

60. What is the chlorine dosage in milligrams per liter if 14.1 mgd is treated with 289 lb/d of chlorine?

The equation is:

$$\textbf{Number of lb/d} = \textbf{(mgd)(Dosage, mg/L)(8.34 lb/gal)}$$

Substituting and rearranging the equation:

$$\text{Chlorine dosage, mg/L} = \frac{289 \text{ lb/d}}{(14.1 \text{ mgd})(8.34 \text{ lb/gal})} = \textbf{2.46 mg/L of chlorine}$$

61. What is the chlorine dosage in milligrams per liter if 38.2 mgd is treated with 956 lb/d of chlorine?

The equation is:

$$\textbf{Number of lb/d} = \textbf{(mgd)(Dosage, mg/L)(8.34 lb/gal)}$$

Substituting and rearranging the equation:

$$\text{Dosage} = \frac{956 \text{ lb/d}}{(38.2 \text{ mgd})(8.34 \text{ lb/gal})}$$

$$\text{Dosage, mg/L} = \textbf{3.00 mg/L of chlorine}$$

62. A water treatment facility is treating 9.6 mgd. If the desired dose for chlorine is 1.2 mg/L, how many pounds of chlorine per day are required?

The equation is:

$$\textbf{Number of lb/d} = \textbf{(mgd)(Dosage, mg/L)(8.34 lb/gal)}$$

$$\text{lb/d} = (9.6 \text{ mgd})(1.2 \text{ mg/L})(8.34 \text{ lb/gal}) = \textbf{96 lb/d of chlorine}$$

63. What is the chlorine dosage in milligrams per liter if 35.2 mgd is treated with 562 lb/d of chlorine?

The equation is:

$$\textbf{Number of lb/d} = \textbf{(mgd)(Dosage, mg/L)(8.34 lb/gal)}$$

$$\frac{(562 \text{ lb/d})}{(35.2 \text{ mgd})(8.34 \text{ lb/gal})} = \text{Dosage}$$

Dosage, mg/L = **1.91 mg/L of chlorine**

64. A water treatment facility is treating 12.9 mgd. If the desired dose for chlorine is 1.75 mg/L, how many pounds of chlorine per day are required?

The equation is:

$$\textbf{Number of lb/day} = \textbf{(mgd)(Dosage, mg/L)(8.34 lb/gal)}$$

lb/d = (12.9 mgd)(1.75 mg/L)(8.34 lb/gal) = **188 lb/d of Cl_2**

65. How many pounds of calcium hypochlorite are needed to treat 20.8 mgd with a 1.80-mg/L dosage if the calcium hypochlorite ($Ca(OCl)_2$) has 63% available chlorine?

The equation is:

$$\textbf{Number of lb/d} = \frac{\textbf{(mgd)(Dosage, mg/L)(8.34 lb/gal)}}{\textbf{\% purity}}$$

$$\text{lb/d} = \frac{(20.8 \text{ mgd})(1.80 \text{ mg/L})(8.34 \text{ lb/gal})}{63\%/100\%}$$

$$\frac{312.2 \text{ lb chlorine required}}{0.63 \text{ available chlorine}} = 495.63 \text{ lb of } Ca(OCl)_2\text{, round to } \mathbf{500 \text{ lb } Ca(OCl)_2}$$

66. Determine the feed rate for alum in milliliters per minute under the following conditions:

Plant flow	**26.2 mgd**
Alum dosage rate	**14.7 mg/L**
Alum percentage	**48.5%**
Alum specific gravity	**1.26**

First, find the pounds per gallon for alum.

$$\textbf{lb/gal alum} = \textbf{(sp gr)(8.34 lb/gal)}$$

$$\text{lb/gal alum} = (1.26)(8.34 \text{ lb/gal}) = 10.51 \text{ lb/gal}$$

The equation for dosage is:

$$\textbf{Dosage, mg/L} = \frac{\textbf{(mL/min)(1,440 min/d)(Alum, lb/gal)(\% purity)}}{\textbf{(mgd)(8.34 lb/gal)(3,785 mL/gal)}}$$

Rearrange to solve for the unknown (milliliters per minute).

$$\text{Alum, mL/min} = \frac{\text{(Dosage, mg/L)(mgd)(8.34 lb/gal)(3,785 mL/gal)}}{\text{(Alum, lb/gal)(\% purity)(1,440 min/d)}}$$

$$\text{Alum, mL/min} = \frac{(14.7 \text{ mg/L})(26.2 \text{ mgd})(8.34 \text{ lb/gal})(3{,}785 \text{ mL/gal})}{(10.51 \text{ lb/gal})(48.5\%/100\% \text{ purity})(1{,}440 \text{ min/d})}$$

$$= \textbf{1,660 mL/min, alum}$$

67. A 2.0-ft diameter pipe that is 3.2 mi long was disinfected with chlorine. If 82.9 lb of chlorine were used, what was the dosage in milligrams per liter?

First, find the number of feet in 3.2 mi.

$$\text{Number of ft} = (5{,}280 \text{ ft/mi})(3.2 \text{ mi}) = 16{,}896 \text{ ft}$$

Next, find the volume in cubic feet for the pipe.

The equation is:

$$\textbf{Volume, ft}^3 = \textbf{(0.785)(Diameter)}^2\textbf{(Length, ft)}$$

$$\text{Volume, ft}^3 = (0.785)(2.0 \text{ ft})(2.0 \text{ ft})(16{,}896 \text{ ft}) = 53{,}053 \text{ ft}^3$$

Then, determine the number of gallons.

$$\text{Number of gal} = (53{,}053 \text{ ft}^3)(7.48 \text{ gal/ft}^3) = 396{,}836 \text{ gal}$$

Convert the number of gallons to million gallons.

$$\text{Million gallons} = \frac{396{,}836 \text{ gal}}{1{,}000{,}000/1 \text{ M}} = 0.396836 \text{ mil gal}$$

Last, find the dosage in milligrams per liter.

$$\textbf{Dosage, mg/L} = \frac{\textbf{lb, chlorine}}{\textbf{(mil gal)(8.34 lb/gal)}}$$

$$\text{Dosage, mg/L} = \frac{82.9 \text{ lb}}{(0.396836 \text{ mil gal})(8.34 \text{ lb/gal})}$$

$$= 25.05 \text{ mg/L, round to } \mathbf{25\ mg/L}$$

68. A treatment plant is adding 245.7 g/min of soda ash to its treated water. If the plant is producing water at 15.25 mgd, what is the soda ash usage in pounds per day and dosage in milligrams per liter?

First, find the pounds per day of soda ash usage.

$$\textbf{Pounds per day of soda ash} = \textbf{(g/min)(1,440 min/d)(1 lb/454 g)} = \textbf{lb/d}$$

$$\text{Pounds per day of soda ash} = (245.7 \text{ g/min})(1{,}440 \text{ min/d})(1 \text{ lb}/454 \text{ g})$$

$$= 779 \text{ lb/d of soda ash}$$

Then, using the "pounds" equation, calculate the dosage in milligrams per liter.

$$\text{Dosage, mg/L} = \frac{779 \text{ lb/d}}{(15.25 \text{ mgd})(8.34 \text{ lb/gal})} = \mathbf{6.12\ mg/L\ of\ soda\ ash}$$

69. A 55-gal drum holds 48 gal of water that has a polymer dosage of 25.79 mg/L. How many ounces of polymer are present in the tank if the polymer is 94.5% pure?

First, solve the problem using the modified "pounds" formula, then convert pounds to ounces.

The equation is:

$$\textbf{Number of lb} = \frac{\textbf{(mil gal)(Dosage, mg/L)(8.34 lb/gal)}}{\textbf{\% purity}}$$

Delete "day" on each side of the equation because it is not required in this problem.

Next, find how many million gallons there are in 48 gal.

$$\text{Million gallons} = \frac{48 \text{ gal}}{1{,}000{,}000/\text{M}} = 0.000048 \text{ mil gal}$$

Substituting:

$$\text{Pounds of polymer} = \frac{(0.000048 \text{ mil gal})(25.79 \text{ mg/L})(8.34 \text{ lb/gal})}{94.5\%/100\%}$$

$$\text{Pounds of polymer} = 0.01 \text{ lb}$$

Now, convert pounds to gallons.

$$\frac{0.01 \text{ lb of polymer}}{8.34 \text{ lb/gal}} = 0.001 \text{ gal}$$

Last, convert gallons to ounces.

Number of oz, polymer = (128 oz/gal)(0.001 gal) = **0.128 oz of polymer**

70. A sodium hypochlorite (NaOCl) solution is going to be prepared in a 55-gal drum. If 5.0 gal of a 12% solution is added to the drum, how much water should be added to make a sodium hypochlorite solution of 2.0%?

The equation is:

$$\textbf{Desired percent NaOCl} = \frac{\textbf{(Hypo, gal)(\% NaOCl)}}{\textbf{Hypo, gal} + \mathbf{H_2O}\ \textbf{needed, gal}}$$

Solve the equation for water needed in gallons. Remember, whatever you do to one side of the equation, you have to do to the other side.

First, multiply both sides of the equation by Hypo, gal + H_2O needed, gal.

This will cancel out the denominator on the right side of the equation giving the following equation:

$$\text{(Hypo, gal} + \text{H}_2\text{O needed, gal)(Desired \%NaOCl)} = \text{(Hypo, gal)(\%NaOCl)}$$

Second, multiply the desired Hypo, % by Hypo, gal + H_2O needed, gal.

$$\text{(Hypo, gal)(Desired \%NaOCl)} + (\text{H}_2\text{O needed, gal})\text{(Desired \%NaOCl)}$$

$$= \text{(Hypo, gal)(\%NaOCl)}$$

Third, subtract (Hypo, gal)(Desired %NaOCl) from both sides of the equation. This cancels out the one on the left.

$$(\text{H}_2\text{O needed, gal})\text{(Desired \%NaOCl)} = \text{(Hypo, gal)(\%NaOCl)}$$

$$- \text{(Hypo, gal)(Desired \%NaOCl)}$$

Fourth, divide both sides by desired %NaOCl. This cancels out the one on the left, leaving the water needed in gallons by itself, which is what needs solving.

$$\text{H}_2\text{O needed, gal} = \frac{\text{(Hypo, gal)(\% NaOCl)} - \text{(Hypo, gal)(Desired \% NaOCl)}}{\text{(Desired \% NaOCl)}}$$

Substitute and solve.

$$\text{H}_2\text{O needed, gal} = \frac{(5.0\text{ gal})(12\%) - (5.0\text{ gal})(2.0\%)}{(2.0\%)} = \frac{60 - 10}{2.0\%} = \frac{50\text{ gal\%}}{2.0\%} = \mathbf{25\ gal}$$

71. Alum is being injected at a rate of 450 mL/min. The alum has a specific gravity of 1.34 and is 48% pure. If this is a dosage of 12.4 mg/L, what is the flow through the plant in million gallons per day?

First, convert the number of milliliters per minute to gallons per day.

$$\frac{(450 \text{ mL/min})(1{,}440 \text{ min/d})}{3{,}785 \text{ mL/gal}} = 171.2 \text{ gpd}$$

Convert specific gravity to pounds per gallon.

$$(1.34)(8.34 \text{ lb/gal}) = 11.176 \text{ lb/d}$$

Next, find the number of pounds per day of alum.

$$(171.2 \text{ gpd})(11.176 \text{ lb/gal}) = 1913.33 \text{ lb/d}$$

Next, find the number of million gallons per day by using the "pounds" formula, but solving for million gallons per day.

Number of lb/d = (mgd)(Dosage, mg/L)(8.34 lb/gal)(% purity)

$$\text{Million gallons per day} = \frac{(\text{Number of lb/d})(\% \text{ purity})}{(\text{Dosage, mg/L})(8.34 \text{ lb/gal})}$$

$$\text{Million gallons per day} = \frac{(1{,}913.33 \text{ lb/d})(48\%/100\%)}{(12.4 \text{ mg/L})(8.34 \text{ lb/gal})} = 8.88 \text{ mgd, round to } \mathbf{8.9 \text{ mgd}}$$

72. How many milliliters per minute of alum are needed if the desired dose is 17.25 mg/L and the plant is treating 20.0 ft³/s? The alum has a specific gravity of 1.29 and is 48% pure.

First, convert cubic feet per second to million gallons per day.

$$\frac{(20.0 \text{ ft}^3\text{/s})(86{,}400 \text{ s/d})(7.48 \text{ gal/ft}^3)}{1{,}000{,}000/1 \text{ mil gal}} = 12.925 \text{ mgd}$$

Next, find the number of pounds per gallon for the alum.

$$\frac{(1.29)(8.34 \text{ lb/gal})(48\% \text{ purity})}{100\%} = 5.16 \text{ lb/gal}$$

Next, find the number of pounds per day of alum using the "pounds" formula.

Alum, lb/d = (mgd)(Dosage, mg/L)(8.34 lb/gal)

Alum, lb/d = (12.925 mgd)(17.25 mg/L)(8.34 lb/gal)

Alum, lb/d = 1,859.46 lb/d

Then, divide the number of pounds per day by the pounds per gallon and number of minutes in a day to get gallons per minute.

$$\frac{(1{,}859.46 \text{ lb/d})}{(5.16 \text{ lb/gal})(1{,}440 \text{ min/d})} = 0.25 \text{ gpm}$$

Last, convert gallons per minute to milliliters per minute.

(3,785 mL/gal)(0.25 gpm) = 946.25 mL/min, round to **950 mL/min of alum**

73. A treatment plant is using zinc orthophosphate (ZOP) for corrosion control. The dosage is 0.28 mg/L. If the plant is treating 19.5 mgd, what should the feeder setting be in milliliters per minute? The specific gravity of the ZOP is 1.52.

First, calculate the pounds per gallon of ZOP.

(1.52 sp gr)(8.34 lb/gal) = 12.68 lb/gal

Next, find the ZOP usage in pounds per day using the "pounds" formula.

Pounds per day = (19.5 mgd)(0.28 mg/L)(8.34 lb/gal) = 45.5 lb/d ZOP

Then, convert pounds per day ZOP to gallons per day.

$$\frac{45.5 \text{ lb/d}}{12.68 \text{ lb/gal}} = 3.59 \text{ gpd ZOP}$$

Last, convert gallons per day ZOP to milliliters per minute.

$$\frac{(3.59 \text{ gpd})(3{,}785 \text{ mL/gal})}{1{,}440 \text{ min/d}} = \mathbf{9.4 \text{ mL/min of ZOP}}$$

74. A 5-min drawdown test result showed that 85 mL of a polymer aid was being used to treat the raw water. The specific gravity of the polymer aid is 1.23. If the plant is treating 2,945 gpm, what is the polymer dosage in milligrams per liter?

First, determine the number of milliliters per minute for the polymer aid (polymer).

$$\frac{85 \text{ mL}}{5 \text{ min}} = 17 \text{ mL/min}$$

Next, calculate the pounds per gallon for the polymer.

$$\textbf{lb/gal} = \textbf{(Specific gravity)(8.34 lb/gal)}$$

$$\text{Polymer, lb/gal} = (1.23)(8.34 \text{ lb/gal}) = 10.26 \text{ lb/gal}$$

Then, find the number of million gallons per day.

$$(2{,}945 \text{ gpm})(1{,}440 \text{ min/d})(1 \text{ M}/1{,}000{,}000) = 4.24 \text{ mgd}$$

Use the dosage equation with conversions added for convenience (dosage/conversion equation).

$$\text{Polymer, mg/L} = \frac{(\text{mL/min})(1{,}440 \text{ min/d})(\text{lb/gal, polymer})}{(3{,}785 \text{ mL/gal})(\text{mgd})(8.34 \text{ lb/gal})}$$

$$\text{Polymer, mg/L} = \frac{(17 \text{ mL/min})(1{,}440 \text{ min/d})(10.26 \text{ lb/gal})}{(3{,}785 \text{ mL/gal})(4.24 \text{ mgd})(8.34 \text{ lb/gal})} = \textbf{1.9 mg/L of polymer aid}$$

75. A dosage of 0.40 mg/L of copper sulfate pentahydrate is needed to control algae in a 10,670-acre-ft reservoir. If the available copper is 25%, how many pounds of copper sulfate pentahydrate are required?

First, convert the volume of the reservoir to million gallons.

$$(43{,}560 \text{ ft}^3/\text{acre-ft})(10{,}670)(1 \text{ M}/1{,}000{,}000) = 464.78 \text{ mil gal}$$

The equation is:

$$\text{Copper sulfate, lb} = \frac{(\text{mil gal})(\text{Dosage, mg/L})(8.34 \text{ lb/gal})}{\text{Percent available copper}/100\%}$$

$$\text{Copper sulfate, lb} = \frac{(464.78)(0.40 \text{ mg/L})(8.34 \text{ lb/gal})}{25\%/100\%} = \textbf{6,200 lb of copper sulfate}$$

76. Calculate the feed rate for fluorosilicic acid in gallons per day given the following data:

Flow rate is 43.5 mgd	**Treated with 21% solution of H_2SiF_6**
Fluoride desired is 1.0 mg/L	**Fluoride ion percent is 79%**
Fluoride in raw water is 0.30 mg/L	**H_2SiF_6 weighs 9.83 lb/gal**

First, determine the required fluoride (F) dose.

$$\textbf{F req.} = \textbf{F desired} - \textbf{F in raw water}$$

$$\text{F req.} = 1.0 \text{ mg/L} - 0.30 \text{ mg/L} = 0.70 \text{ mg/L}$$

Next, calculate the number of pounds of fluoride needed using the pounds formula.

$$\text{F, lb} = (\text{mg/L})(\text{mgd})(8.34 \text{ lb/gal})$$

$$\text{F, lb} = (0.70 \text{ mg/L})(43.5 \text{ mgd})(8.34 \text{ lb/gal}) = 253.95 \text{ lb (F)}$$

Next, determine the pounds of 79% fluoride.

$$\frac{253.95 \text{ lb (F)}}{79\%/100\%} = 321.46 \text{ lb of } 79\% \text{ F}$$

$$\frac{321.46 \text{ lb}}{21\%/100\%} = 1{,}531 \text{ lb of } 21\% \text{ F solution}$$

$$\text{Number of gal } H_2SiF_6 = \frac{1{,}531 \text{ lb of } 21\% \text{ solution}}{9.83 \text{ lb/gal}} = 155.7 \text{ gpd, round to } \mathbf{160 \text{ gpd } H_2SiF_6}$$

Another way to solve this problem follows:

First, determine the required fluoride dose.

$$\textbf{F req.} = \textbf{F desired} - \textbf{F in raw water}$$

$$\text{F req.} = 1.0 \text{ mg/L} - 0.30 \text{ mg/L} = 0.70 \text{ mg/L}$$

Next, calculate the number of milligrams per liter of H_2SiF_6.

$$\text{Number of mg/L of } H_2SiF_6 = \frac{(\% \text{ solution})(10{,}000 \text{ mg/L})}{1\%} = \frac{(21\% \text{ F})(10{,}000 \text{ mg/L})}{1\%} =$$

$$\text{Number of mg/L of } H_2SiF_6 = 210{,}000 \text{ mg/L}$$

Now, solve for the million gallons per day flow of fluoride.

(continued)

The equation is:

(F, mg/L)(mgd)(8.34 lb/gal) = (% as mg/L)(mgd flow)(Solution, lb/gal)(% F ion)

$$\%F = 79\%/100\% = 0.79\ F$$

$$(0.7\ \text{mg/L F})(43.5\ \text{mgd})(8.34\ \text{lb/gal}) = (210{,}000\ \text{mg/L})(x\ \text{mgd flow})(9.83\ \text{lb/gal})(0.79\ \text{F})$$

Solve for x.

$$x\ \text{mgd flow} = \frac{(0.70\ \text{mg/L F})(43.5\ \text{mgd})(8.34\ \text{lb/gal})}{(210{,}000\ \text{mg/L})(9.83\ \text{lb/gal})(0.79\ \text{F})} = 0.0001557\ \text{mgd}$$

Convert million gallons per day to gallons.

(0.0001557 mgd)(1,000,000 gal/1 M) = 155.7 gpd, round to **160 gpd H_2SiF_6**

77. How many fluid ounces of sodium hypochlorite (5.25% available chlorine) are required to disinfect a well with the following parameters: (1) depth of well is 316 ft; (2) 12-in. diameter well casing extends down to 100.0 ft; (3) the remainder is a 10.0-in. diameter casing; (4) the residual desired dose is 50.0 mg/L; (5) the depth to water is 83.5 ft; and (6) the chlorine demand is 6.0 mg/L.

First, find the diameter in feet for both well casings.

$$\text{Diameter for 12.0-in. casing} = \frac{(12.0\ \text{in.})(1\ \text{ft})}{12\ \text{in.}} = 1.00\ \text{ft}$$

$$\text{Diameter for 10.0-in. casing} = \frac{(10.0\ \text{in.})(1\ \text{ft})}{12\ \text{in.}} = 0.833\ \text{ft}$$

Then, find the length (in feet) of water in the casing.

Length of water-filled casing = Depth of well – Depth of water to top of casing

Length of water-filled casing for 12.0-in. diameter = 100 ft – 83.5 ft = 16.5 ft

Length of water-filled casing for 10.0-in. diameter = 316 ft – 100 ft = 216 ft

Next, determine the volume (in gallons) of water in the well casings using the following formula:

$$\textbf{Volume, gal} = \mathbf{(0.785)(Diameter)^2(Length)(7.48\ gal/ft^3)}$$

$$\text{Volume, 12.0-in. casing} = (0.785)(1.00\text{ ft})(1.00\text{ ft})(16.5\text{ ft})(7.48\text{ gal/ft}^3) = 97\text{ gal}$$

$$\text{Volume, 10.0-in. casing} = (0.785)(0.833\text{ ft})(0.833\text{ ft})(216\text{ ft})(7.48\text{ gal/ft}^3) = \underline{880\text{ gal}}$$

$$\text{Total water in casings} = 977\text{ gal}$$

Next, determine the number of million gallons.

$$\text{Million gallons} = (977\text{ gal})(1\text{ M}/1{,}000{,}000) = 0.000977\text{ mil gal}$$

Then, find the chlorine required.

$$\text{Chlorine required} = \text{Chlorine demand} + \text{Chlorine residual}$$

$$Cl_2\text{ req.} = 6.0\text{ mg/L} + 50.0\text{ mg/L} = 56\text{ mg/L}$$

Then, using the "pounds" equation, calculate the number of pounds of sodium hypochlorite.

$$\text{Gallons of sodium hypochlorite} = \frac{(0.000971\text{ mil gal})(56\text{ mg/L})}{5.25\%/100\%\text{ Available chlorine}}$$

$$\text{Gallons of sodium hypochlorite} = 1.036\text{ gal}$$

Last, convert to fluid ounces.

$$1.036\text{ gal}(128\text{ oz/gal}) = 132.6\text{ oz, round to } \mathbf{130\ oz}$$

78. A water treatment plant has a filter flow of 6,500 gpm and is being treated with 850 gpd of a hypochlorite solution. The desired dose is 2.0 mg/L; determine the concentration of the hypochlorite solution, in percent.

First, convert 6,500 gpm filter flow to million gallons per day.

$$\frac{(6{,}500 \text{ gpm})\ (1{,}440 \text{ min})\ \ (1 \text{ mil gal})}{\text{day} \quad\quad 1{,}000{,}000 \text{ gal}} = 9.36 \text{ mgd}$$

Then, convert 850 gpd of the hypochlorite solution rate of flow to million gallons per day.

$$(850 \text{ gpd}) \frac{(1 \text{ mil gal})}{1{,}000{,}000 \text{ gal}} = 0.00085 \text{ mgd}$$

Then, using the equal dosage equations:

$$(0.00085 \text{ mgd})(x \text{ mg/L})(8.34 \text{ lb/gal}) = (9.36 \text{ mgd})(2.0 \text{ mg/L})(8.34 \text{ lb/gal})$$

$$\text{Milligrams per liter, hypochlorite} = \frac{(9.36 \text{ mgd})(2.0 \text{ mg/L})(8.34 \text{ lb/gal})}{(0.00085 \text{ mgd})(8.34 \text{ lb/gal})} = 22{,}024 \text{ mg/L}$$

Last, convert milligrams per liter to percent.

$$(22{,}024 \text{ mg/L}) \frac{(1\%)}{10{,}000 \text{ mg/L}} = \mathbf{2.2\%}$$

79. A treatment plant is adding 341.65 g/min of soda ash to its treated water. If the plant is producing water at a rate of 15 ft³/s, what is the soda ash usage in pounds per day and dosage in milligrams per liter?

First, find the pounds per day of soda ash usage.

The equation is:

Pounds per day of soda ash = (g/min)(1,440 min/d)(1 lb/454 g)

Pounds per day of soda ash = (341.65 g/min)(1,440 min/d)(1 lb/454 g)
= **1,084 lb/d of soda ash**

Next, convert 15 ft^3/s to the number of million gallons per day.

Number of mgd = (15 ft^3/s)(86,400 s/d)(7.48 gal/ft^3)(1 M/1,000,000) = 9.694 mgd

Then, rearranging the "pounds" equation to solve for dosage, calculate the dosage in milligrams per liter.

$$\text{Dosage, mg/L} = \frac{1{,}084 \text{ lb/d}}{(9.694 \text{ mgd})(8.34 \text{ lb/gal})} = 13.408 \text{ mg/L, round to } \mathbf{13\ mg/L\ soda\ ash}$$

POUNDS REMOVAL PROBLEMS

80. Find the amount of iron (Fe) and manganese (Mn) removed per year from a water plant that treats an average of 30.4 mgd of source water if the average iron concentration is 1.54 mg/L and the average manganese concentration is 0.21 mg/L. The removal efficiency is 86.8% for the iron and 72.9% for the manganese.

First, calculate the iron and manganese removal in milligrams per liter.

$$(1.54 \text{ mg/L})(86.8\%/100\%) = (1.54 \text{ mg/L})(0.868) = 1.337 \text{ mg/L}$$

$$(0.21 \text{ mg/L})(72.9\%/100\%) = (0.21 \text{ mg/L})(0.729) = 0.153 \text{ mg/L}$$

Determine the amount of water in million gallons produced for the year.

$$(30.4 \text{ mgd})(365 \text{ d/yr}) = 11{,}096 \text{ mil gal/yr}$$

Next, using the "pounds" equation, solve for the number of pounds per year for iron and manganese.

Pounds per year = (mil gal/yr)(Dosage, mg/L)(8.34 lb/gal)

Pounds per year, Fe removal = (11,096 mil gal/yr)(1.337 mg/L)(8.34 lb/gal)

= 123,727 lb/yr, round to **124,000 lb/yr of Fe removed**

Pounds per year, Mn removal = (11,096 mil gal/yr)(0.153 mg/L)(8.34 lb/gal)

= 14,159 lb/yr, round to **14,000 lb/yr of Mn removed**

81. Find the amount of iron and manganese removed per year from a plant that treats an average of 61.4 mgd if the average iron concentration is 1.08 ppm and the average manganese concentration is 0.19 ppm. The removal efficiency is 88.2% for the iron and 69.4% for the manganese.

First, calculate the iron and manganese removal in parts per million.

$$(1.08 \text{ ppm})(88.2\%/100\%) = 1.08 \text{ ppm}(0.882) = 0.953 \text{ ppm}$$

$$(0.19 \text{ ppm})(69.4\%/100\%) = 0.19 \text{ ppm}(0.694) = 0.132 \text{ ppm}$$

Determine the amount of water in million gallons produced for the year.

$$(61.4 \text{ mgd})(365 \text{ d/yr}) = 22{,}411 \text{ mil gal/yr}$$

Next, using the "pounds" equation, solve for the number of pounds per year for iron and manganese.

$$\text{Pounds per year} = (\text{mil gal/yr})(\text{Dosage, mg/L})(8.34 \text{ lb/gal})$$

$$\text{Pounds per year, Fe removal} = (22{,}411 \text{ mil gal/yr})(0.953 \text{ mg/L})(8.34 \text{ lb/gal})$$

= 178,123 lb/yr, round to **178,000 lb/yr of Fe removed**

$$\text{Pounds per year, Mn removal} = (22{,}411 \text{ mil gal/yr})(0.132 \text{ mg/L})(8.34 \text{ lb/gal})$$

= 24,672 lb/yr, round to **25,000 lb/yr of Mn removed**

PUMP DISCHARGE PROBLEMS

82. How many gallons will a pump discharge if it pumps an average of 25 gpm for 2 hr and 25 min?

First, calculate the number of minutes for the time interval given.

$$2 \text{ hr}(60 \text{ min/hr}) + 25 \text{ min} = 120 \text{ min} + 25 \text{ min} = 145 \text{ min}$$

Then, calculate the number of gallons by multiplying the length of time the pump worked in minutes times the number of gallons per minute.

145 min(25 gpm) = 3,625 gal, round to **3,600 gal**

83. How long will it take in hours and minutes to unload a tanker truck filled with liquid alum if the truck's pump unloads the alum at 90 gpm and a total of 14,500 L needs to be unloaded? The storage tank's capacity is 10,000 gal and it has about 2,000 gal of alum in it.

First, determine the number of gallons in 14,500 L.

$$\text{Number of gal} = \frac{14{,}500 \text{ L}}{3.785 \text{ L/gal}} = 3{,}831 \text{ gal}$$

Then, divide the number of gallons by the pumping rate.

$$\text{Time to pump} = 3{,}831 \text{ gal/90 gpm} = 42.57 \text{ min, round to 43 min}$$

The unloading time will be 43 min.

84. A pump's output is averaging 58 gpm. How many gallons will it pump in one day?

(58 gpm)(1,440 min/d) = 83,520 gpd, round to **84,000 gpd**

85. How long will it take in hours for a pump to discharge 25,000 gal if it is pumping at a rate of 43 gpm?

First, determine the number of minutes it will take.

$$\frac{(25{,}000 \text{ gal})}{(43 \text{ gal/min})} = 581.4 \text{ min}$$

Then, convert to hours by dividing by 60 min/hour.

$$\frac{581.4 \text{ min}}{60 \text{ min/hr}} = 9.69 \text{ hr, round to } \mathbf{9.7 \text{ hr}}$$

86. How long will it take in hours and minutes to empty a tanker truck with ZOP if the truck's pump unloads the ZOP at 58 gpm and a total of 18,800 L needs to be unloaded? The storage tank's capacity is 10,000 gal and it has about 1,000 gal of ZOP.

First, determine the number of gallons in 18,800 L.

$$\text{Number of gal} = \frac{18{,}800 \text{ L}}{3.785 \text{ L/gal}} = 4{,}967 \text{ gal}$$

Then, divide the number of gallons by the pumping rate.

$$\text{Time to pump} = 4{,}967 \text{ gal/58 gpm} = 85.64 \text{ min}$$

Divide by 60 min/hr.

$$85.64 \text{ min/(60 min/hr)} = 1.43 \text{ hr}$$

Next, determine how many minutes are in 0.43 hr by multiplying by 60 min/hr.

$$(0.43 \text{ hr})(60 \text{ min/hr}) = 25.8 \text{ min, round to 26 min}$$

The unloading time will be 1 hr and 26 min.

PUMPING CALCULATIONS

87. Water is being pumped from a water source with an elevation of 3,452 ft to an elevation of 3,609 ft. What is the total head if friction and minor head losses are 17 ft?

The equation is:

Total head, ft = Total static head, ft (difference in elevation) + Head losses, ft

Total head, ft = (3,609 ft – 3,452 ft) + 17 ft = 157 + 17 ft = **174 ft**

88. Find the motor horsepower for a pump station with the following parameters:

Motor efficiency:	**90%**	**Total head:**	**186 ft**
Pump efficiency:	**79%**	**Flow:**	**1.75 mgd**

First, convert million gallons per day to gallons per minute.

Gallons per minute = (1.75 mgd)(1,000,000/1 M)(1 d/1,440 min) = 1,215 gpm

Next, calculate the motor horsepower.

The equation is:

$$\textbf{Motor horsepower} = \frac{\textbf{(Flow, gpm)(Total head, ft)}}{\textbf{(3,960)(Motor efficiency)(Pump efficiency)}}$$

$$\frac{(1{,}215 \text{ gpm})(186 \text{ ft})}{(3{,}960)(90\%/100\% \text{ Motor efficiency})(79\%/100\% \text{ Pump efficiency})} = \textbf{80 mhp}$$

89. What is the brake horsepower if the water horsepower is 26 and the pump efficiency is 86%?

The equation is:

Brake horsepower = Water horsepower/Pump efficiency

Brake horsepower = 26 whp/(86%/100% Pump efficiency)

= 30.2 bhp, round to **30 bhp**

90. Find the water horsepower for a system with the following characteristics: Motor efficiency is 91%; pump efficiency is 75%; and motor horsepower is 125.

The equation is:

Water horsepower = (mhp)(Motor efficiency)(Pump efficiency)

Water horsepower = (125 mhp)(91%/100% Motor efficiency)
(75%/100 Pump efficiency)
= 85.31 whp, round to **85 whp**

91. Find the water horsepower for a system with the following characteristics: Motor efficiency is 95%; pump efficiency is 86%; and motor horsepower is 100.

The equation is:

Water horsepower = (mhp)(Motor efficiency)(Pump efficiency)

Water horsepower = (100 mhp)(95%/100% Motor efficiency)
(86%/100% Pump efficiency)
= 81.7 whp, round to **80 whp**

92. What is the motor horsepower if the brake horsepower is 55 and the motor efficiency is 90%?

The equation is:

Motor horsepower = Brake horsepower/Motor efficiency

Motor horsepower = 55 bhp/(90%/100% Motor efficiency)
= 61.1 mhp, round to **60 mhp**

93. Find the water horsepower if the brake horsepower is 62 and the pump efficiency is 85%.

The equation is:

$$\textbf{Water horsepower} = \textbf{(Brake horsepower)(Pump efficiency)}$$

$$\text{Water horsepower} = (62\text{ bhp})(85\%/100\%\text{ Pump efficiency})$$

$$= 52.7\text{ whp, round to } \mathbf{53\ whp}$$

94. What is the motor horsepower if 75 hp is required to run a pump with a motor efficiency of 85% and a pump efficiency of 78%? Note: The 75 hp in this problem is called the water horsepower. The water horsepower is the actual energy (horsepower) available to pump water.

The equation is:

$$\textbf{Motor horsepower} = \frac{\textbf{(whp)}}{\textbf{(Motor efficiency)(Pump efficiency)}}$$

$$\text{Motor horsepower} = \frac{(75\text{ whp})}{(85\%/100\%\text{ Motor efficiency})(78\%/100\%\text{ Pump efficiency})}$$

$$\text{Motor horsepower} = \frac{(75\text{ whp})}{(0.85\text{ Motor efficiency})(0.78\text{ Pump efficiency})}$$

$$\text{Motor horsepower} = 113\text{ mhp, round to } \mathbf{110\ mhp}$$

FLOW RATE PROBLEMS

95. What is the flow in gallons per minute from a faucet that fills a 5.0-gal container in 2 min and 23 s?

First, convert the number of seconds to minutes.

$$\frac{23 \text{ s}}{60 \text{ s/min}} = 0.38 \text{ min}$$

Thus, the total time is 2.38 min.

The equation is:

Flow = Volume/Time

Flow = 5.0 gal/2.38 min = **2.1 gpm**

96. Water is flowing through a channel that is 4.0 ft wide and 1.5 ft deep. If the flow is 10.5 ft^3/s, what is the velocity of the water?

The equation is:

Q (Flow) = (Area)(Velocity)

Substituting:

$$10.5 \text{ ft}^3\text{/s} = (4.0 \text{ ft})(1.5 \text{ ft})(\text{Velocity})$$

Solve for velocity by dividing both sides by the area (4 ft × 1.5 ft).

$$\text{Velocity} = \frac{10.5 \text{ ft}^3\text{/s}}{(4.0 \text{ ft})(1.5 \text{ ft})} = 1.75 \text{ ft/s, round to } \mathbf{1.8 \text{ ft/s}}$$

97. Flow through a 5.0-ft-wide channel is 13.4 ft^3/s. If the velocity is 2.2 ft/s, what is the depth of the water in the channel?

The equation is:

Q (Flow) = (Area)(Velocity)

$$13.4 \text{ ft}^3\text{/s} = (5.0 \text{ ft})(x \text{ ft, depth})(2.2 \text{ ft/s})$$

$$x \text{ ft, depth} = \frac{13.4 \text{ ft}^3\text{/s}}{(5.0 \text{ ft})(2.2 \text{ ft})} = \mathbf{1.2 \text{ ft depth}}$$

98. Water is flowing at a velocity of 1.08 ft/s in a 12.0-in. diameter pipe. If the pipe changes from 12.0 in. to 6.0 in., what will the velocity be in the 6-in. pipe?

***Note*:** Water flowing from a smaller diameter pipe to a larger diameter pipe will slow down. Conversely, water flowing from a larger diameter pipe into a smaller diameter pipe (as in this problem) will speed up. When solving problems like this, use this as a cross check to the math.

Flow in the 12.0-in. pipe equals the flow in the 6.0-in. pipe because the flow must remain constant.

$$\mathbf{Q_1 = Q_2}$$

Write the equation as above.

$$\textbf{(Area 1)(Velocity 1)} = \textbf{(Area 2)(Velocity 2)}$$

First, find the diameters for the 6.0-in. and 12.0-in. pipes.

$$\text{Diameter for 6.0-in.} = (\text{6.0-in.})(1 \text{ ft}/12 \text{ in.}) = 0.5 \text{ ft}$$

$$\text{Diameter for 12.0-in.} = (\text{12.0-in.})(1 \text{ ft}/12 \text{ in.}) = 1.0 \text{ ft}$$

Then, determine the areas of each pipe.

$$\text{Area} = (0.785)(\text{Diameter})^2$$

$$\text{Area 1 (6.0-in.)} = (0.785)(0.5 \text{ ft})(0.5 \text{ ft}) = 0.196 \text{ ft}^2$$

$$\text{Area 2 (12.0-in.)} = (0.785)(1.0 \text{ ft})(1.0 \text{ ft}) = 0.785 \text{ ft}^2$$

Last, substitute areas calculated and known velocity in 12-in. pipe.

$$(0.196 \text{ ft}^2)(x, \text{ ft/s}) = (0.785 \text{ ft}^2)(1.25 \text{ ft/s})$$

Solve for x.

$$x, \text{ ft/s} = \frac{(0.785 \text{ ft}^2)(1.08 \text{ ft/s})}{(0.196 \text{ ft}^2)} = 4.32 \text{ ft/s, round to } \textbf{4.3 ft/s in the 6.0-in. pipe}$$

Note that this is not twice as much, because the flow is based on the area of the pipes and only indirectly on the diameter.

WEIR OVERFLOW RATE PROBLEMS

99. A rectangular clarifier has a weir length of 95.5 ft. What is the weir overflow rate in gallons per day per foot if the flow is 1.45 mgd?

The equation is:

$$\textbf{Weir overflow rate} = \frac{\textbf{Flow, gpd}}{\textbf{Weir length, ft}}$$

Convert 1.45 mgd to gallons per day.

$$1.45 \text{ mgd}(1{,}000{,}000/1 \text{ M}) = 1{,}450{,}000 \text{ gpd}$$

$$\text{Weir overflow rate} = \frac{1{,}450{,}000 \text{ gpd}}{95.5 \text{ ft}} = 15{,}183\text{, round to } \mathbf{15{,}200 \text{ gpd/ft}}$$

100. A circular clarifier has a weir length of 155 ft. What is the weir overflow rate in gallons per day per foot if the flow is 3.08 mgd?

The equation is:

$$\textbf{Weir overflow rate} = \frac{\textbf{Flow, gpd}}{\textbf{Weir length, ft}}$$

Convert 3.08 mgd to gallons per day.

$$3.08 \text{ mgd}(1{,}000{,}000/1 \text{ M}) = 3{,}080{,}000 \text{ gpd}$$

$$\text{Weir overflow rate} = \frac{3{,}080{,}000 \text{ gpd}}{155 \text{ ft}} = 19{,}871 \text{ gpd/ft, round to } \mathbf{19{,}900 \text{ gpd/ft}}$$

101. A circular clarifier has a weir length of 185 ft. What is the weir overflow rate in gallons per day per foot if the flow is 7.15 mgd?

The equation is:

$$\textbf{Weir overflow rate} = \frac{\textbf{Flow, gpd}}{\textbf{Weir length, ft}}$$

Convert 7.15 mgd to gallons per day.

$$7.15 \text{ mgd}(1{,}000{,}000/1 \text{ M}) = 7{,}150{,}000 \text{ gpd}$$

$$\text{Weir overflow rate} = \frac{7{,}150{,}000 \text{ gpd}}{185 \text{ ft}} = 38{,}649 \text{ gpd/ft, round to } \textbf{38,600 gpd/ft}$$

SURFACE LOADING RATE CALCULATIONS

102. What is the surface loading rate for a sedimentation basin that is 350 ft by 65 ft if it is treating an instantaneous flow rate of 14 ft³/s?

First, convert the number of cubic feet per second to gallons per day.

$$(14 \text{ ft}^3/\text{s})(86{,}400 \text{ s/d})(7.48 \text{ gal/ft}^3) = 9{,}047{,}808 \text{ gpd}$$

The equation for surface loading rate is:

$$\textbf{Surface loading rate} = \frac{\textbf{gpd}}{\textbf{Number of ft}^2}$$

$$\text{Surface loading rate} = \frac{9{,}047{,}808 \text{ gpd}}{(350 \text{ ft})(65 \text{ ft})} = 397.7 \text{ gpd/ft}^2\text{, round to } \textbf{400 gpd/ft}^2$$

103. What is the surface loading rate for a sedimentation basin that is 265 ft by 61 ft if it is treating an instantaneous flow rate of 13.4 ft^3/s?

First, convert the number of cubic feet per second to gallons per day.

$$(13.4 \text{ ft}^3/\text{s})(86{,}400 \text{ s/d})(7.48 \text{ gal/ft}^3) = 8{,}660{,}045 \text{ gpd}$$

The equation for surface loading rate is:

$$\textbf{Surface loading rate} = \frac{\textbf{gpd}}{\textbf{Number of ft}^2}$$

$$\text{Surface loading rate} = \frac{8{,}660{,}045 \text{ gpd}}{(265 \text{ ft})(61 \text{ ft})} = 535.7 \text{ gpd/ft}^2\text{, round to } \mathbf{540 \text{ gpd/ft}^2}$$

FILTRATION RATE PROBLEMS

104. A filter is 25 ft by 32 ft. What is the filtration rate in gallons per minute if the filter receives a flow of 3,000 gpm?

First, find the surface area of the filter.

$$\text{Filter surface area} = (25 \text{ ft})(32 \text{ ft}) = 800 \text{ ft}^2$$

The equation for filtration rate is:

$$\textbf{Filtration rate} = \frac{\textbf{Flow rate, gpm}}{\textbf{Filter surface area, ft}^2}$$

$$\text{Filtration rate} = \frac{3{,}000 \text{ gpm}}{800 \text{ ft}^2} = 3.75 \text{ gpm/ft}^2\text{, round to } \mathbf{4 \text{ gpm/ft}^2}$$

105. A filter has a surface area of 525 ft. What is the filtration rate in gallons per minute if the filter receives a flow of 3,120 gpm?

$$\textbf{Filtration rate} = \frac{\textbf{Flow rate, gpm}}{\textbf{Filter surface area, ft}^2}$$

$$\text{Filtration rate} = \frac{3{,}120 \text{ gpm}}{525 \text{ ft}^2} = \mathbf{5.94\ gpm/ft^2}$$

BACKWASH RATE CALCULATIONS

106. What is the backwash rate for a filter that has a surface area of 320 ft² and a backwash flow of 3,820 gpm in gallons per minute per square foot?

The equation is:

$$\textbf{Backwash rate, gpm/ft}^2 = \frac{\textbf{Flow, gpm}}{\textbf{Filter area, ft}^2}$$

$$\text{Backwash rate, gpm/ft}^2 = \frac{3{,}820 \text{ gpm}}{320 \text{ ft}^2} = 11.94 \text{ gpm/ft}^2\text{, round to } \mathbf{12\ gpm/ft^2}$$

107. What is the backwash rate in gallons per minute per square foot given the following:

Filter is 18 ft long and 15 ft wide

Backwash flow is 12 ft³/s

First, calculate the area of the filter in square feet.

The equation is:

$$\textbf{Number of ft}^2 = \textbf{(Length, ft)(Width, ft)}$$

$$\text{Number of ft}^2 = (18 \text{ ft})(15 \text{ ft}) = 270 \text{ ft}^2$$

(continued)

Next, convert cubic feet per second to gallons per minute.

The equation is:

$$\textbf{Number of gpm} = \mathbf{(ft^3/s)(60\ s/min)(7.48\ gal/ft^3)}$$

$$\text{Number of gpm} = (12\ \text{ft}^3/\text{s})(60\ \text{s/min})(7.48\ \text{gal/ft}^3) = 5{,}385.6\ \text{gpm}$$

Then, determine backwash rate.

The equation is:

$$\textbf{Backwash rate, gpm/ft}^2 = \frac{\textbf{Flow, gpm}}{\textbf{Filter area, ft}^2}$$

$$\text{Backwash rate, gpm/ft}^2 = \frac{5{,}385.6\ \text{gpm}}{270\ \text{ft}^2}$$

$$\text{Backwash rate, gpm/ft}^2 = 19.95\ \text{gpm/ft}^2\text{, round to } \mathbf{20\ gpm/ft^2}$$

UNIT FILTER RUN VOLUME DETERMINATIONS

108. A filter produced a total of 1,965,100 gal between backwashes. If the filter is 20 ft by 20 ft (measured to nearest foot), what is the unit filter run volume (UFRV) in gallons per square foot?

$$\textbf{UFRV, gal/ft}^2 = \frac{\textbf{Total gallons filtered}}{\textbf{Filter surface area, ft}^2}$$

$$\text{UFRV, gal/ft}^2 = \frac{1{,}965{,}100\ \text{gal}}{(20\ \text{ft})(20\ \text{ft})} = 4{,}913\ \text{gal/ft}^2\text{, round to } \mathbf{4{,}900\ gal/ft^2}$$

109. A filter produced a total of 3,178,500 gal between backwashes. If the filter is 30.0 ft by 25 ft, what is the UFRV in gallons per square foot?

$$\textbf{UFRV, gal/ft}^2 = \frac{\textbf{Total gallons filtered}}{\textbf{Filter surface area, ft}^2}$$

$$\text{UFRV, gal/ft}^2 = \frac{3{,}178{,}500 \text{ gal}}{(30.0 \text{ ft})(25 \text{ ft})} = 4{,}238 \text{ gal/ft}^2\text{, round to } \mathbf{4{,}200 \text{ gal/ft}^2}$$

110. A filter produced a total of 3,095,000 gal between backwashes. If the filter is 28 ft by 32 ft, what is the UFRV in gallons per square foot?

$$\textbf{UFRV, gal/ft}^2 = \frac{\textbf{Total gallons filtered}}{\textbf{Filter surface area, ft}^2}$$

$$\text{UFRV, gal/ft}^2 = \frac{3{,}095{,}000 \text{ gal}}{(28 \text{ ft})(32 \text{ ft})} = 3{,}454 \text{ gal/ft}^2\text{, round to } \mathbf{3{,}500 \text{ gal/ft}^2}$$

CALCULATIONS FOR WELLS

111. Calculate the well yield in gallons per minute, given a drawdown of 8.2 ft and a specific yield of 48 gpm/ft.

Well yield, gpm = (Specific yield, gpm/ft)(Drawdown, ft)

Substituting:

Well yield, gpm = (48 gpm/ft)(8.2 ft) = 393.6 gpm, round to **390 gpm**

112. Find the specific yield in gallons per minute per foot if a well produces 46 gpm and the drawdown for the well is 17 ft.

The equation is:

$$\textbf{Specific yield, gpm/ft} = \frac{\textbf{Well yield, gpm}}{\textbf{Drawdown, ft}}$$

$$\text{Specific yield, gpm/ft} = \frac{46 \text{ gpm}}{17 \text{ ft}} = \textbf{2.7 gpm/ft}$$

113. Find the specific yield in gallons per minute per foot if a well produces 80 gpm and the drawdown for the well is 24.5 ft.

The equation is:

$$\textbf{Specific yield, gpm/ft} = \frac{\textbf{Well yield, gpm}}{\textbf{Drawdown, ft}}$$

$$\text{Specific yield, gpm/ft} = \frac{80 \text{ gpm}}{24.5 \text{ ft}} = 3.27 \text{ gpm/ft, round to } \textbf{3 gpm/ft}$$

114. What is the drawdown for a well that has a static level of 23.45 ft and a pumping water level of 46.8 ft?

The equation is:

$$\textbf{Drawdown, ft} = \textbf{Pumping water level, ft} - \textbf{Static water level, ft}$$

$$\text{Drawdown, ft} = 46.8 \text{ ft} - 23.45 \text{ ft} = \textbf{23.4 ft}$$

115. If the static level in the well was 104.5 ft and the drawdown was 11.33 ft, what was the pumping water level in the well?

The equation is:

Drawdown, ft = Pumping water level, ft – Static water level, ft

Rearrange the equation to solve for pumping water level.

Pumping water level, ft = Drawdown, ft + Static water level, ft

Substitute known values.

Pumping water level, ft = 11.33 ft + 104.5 ft = 115.83 ft, round to **115.8 ft**

116. The pumping water level in the well was 152.07 ft. If the drawdown was 9.01 ft, what was the static water level in the well?

The equation is:

Drawdown, ft = Pumping water level, ft – Static water level, ft

Rearrange the equation to solve for static water level.

Static water level, ft = Pumping water level, ft – Drawdown, ft

Substitute known values.

Static water level, ft = 152.07 ft – 9.01 ft = **143.06 ft**

HARDNESS PROBLEMS

117. A water sample contains 98 mg/L of calcium and 41 mg/L of magnesium. What is the total hardness as $CaCO_3$?

The equation is:

$$\textbf{Total hardness, mg/L as CaCO}_3 = \textbf{Ca hardness, mg/L as CaCO}_3 + \textbf{Mg hardness, mg/L as CaCO}_3$$

Substituting:

$$\text{Total hardness, mg/L as CaCO}_3 = 98 \text{ mg/L Ca} + 41 \text{ mg/L Mg}$$

$$= 139 \text{ mg/L, round to } \textbf{140 mg/L as CaCO}_3$$

118. A water sample contains 102 mg/L of calcium and 36 mg/L of magnesium. What is the total hardness as $CaCO_3$?

The equation is:

$$\textbf{Total hardness, mg/L as CaCO}_3 = \textbf{Ca hardness, mg/L as CaCO}_3 + \textbf{Mg hardness, mg/L as CaCO}_3$$

$$\text{Total hardness, mg/L as CaCO}_3 = 102 \text{ mg/L Ca} + 36 \text{ mg/L Mg}$$

$$= 138 \text{ mg/L, round to } \textbf{140 mg/L as CaCO}_3$$

119. The calcium content of a water sample is 17 mg/L. What is the calcium hardness as milligrams per liter $CaCO_3$?

The equation is:

$$\textbf{Calcium hardness as mg/L CaCO}_3 = \textbf{2.5(Calcium content in mg/L)}$$

$$\text{Calcium hardness as mg/L CaCO}_3 = 2.5(17 \text{ mg/L calcium})$$

$$\text{Calcium hardness as mg/L CaCO}_3 = 42.5 \text{ mg/L, round to } \textbf{43 mg/L as CaCO}_3$$

120. The magnesium content of a water sample is 8.4 mg/L. What is the calcium hardness expressed as milligrams per liter $CaCO_3$?

The equation is:

Calcium hardness as mg/L $CaCO_3$ = 4.1(Magnesium content in mg/L)

Calcium hardness as mg/L $CaCO_3$ = 4.1(8.4 mg/L magnesium)

Calcium hardness as mg/L $CaCO_3$ = 34.44 mg/L, round to **34 mg/L as $CaCO_3$**

SOLUTION, PERCENT SOLUTION, AND MIXTURE PROBLEMS

121. How many gallons of an 8% solution must be mixed with a 1.3% solution to make 250 gal of a 3% solution?

Solve the problem using the dilution triangle.

8%   1.7   1.7 parts of the 8% solution are required for every 6.7 parts

    3%

1.3%   5   5 parts of the 1.3% solution are required for every 6.7 parts

    6.7 parts

$$\frac{1.7 \text{ parts } (250 \text{ gal})}{6.7 \text{ total parts}} = 63 \text{ gal, round to } \mathbf{60 \text{ gal of the } 8\% \text{ solution}}$$

$$\frac{5 \text{ parts } (250 \text{ gal})}{6.7 \text{ parts}} = \frac{187 \text{ gal, round to } \mathbf{190 \text{ gal of the } 1.3\% \text{ solution}}}{250 \text{ gal}}$$

To make 250 gal of the 3% solution, mix 60 gal of the 8% solution with 190 gal of the 1.3% solution.

122. How many pounds of high test hypochlorite (HTH) are needed to make exactly 200 gal of a 7.0% HTH solution?

The equation is:

$$\textbf{Percent HTH solution} = \frac{\textbf{(lb HTH)(100\%)}}{\textbf{(Number of gal)(8.34 lb/gal)}}$$

Rearranging the equation:

Number of pounds of HTH = (% solution)(Number of gal)(8.34 lb/gal)/100%

Number of pounds of HTH = (7.0% solution)(200 gal)(8.34 lb/gal)/100%

Number of pounds of HTH = 117 lb of HTH, round to **120 lb of HTH**

123. How many pounds of 7.8% sodium hypochlorite are there in 1 gal if the solution weighs 9.24 lb/gal?

$$\text{Number of lb/gal} = \frac{\text{(Solution \%)(8.34 lb/gal)}}{100\%}$$

$$\text{Number of lb/gal} = \frac{\text{(7.8\%)(8.34 lb/gal)}}{100\%} = \textbf{0.65 lb/gal}$$

KILOWATT DETERMINATIONS

124. How many kilowatts will it take to operate a 200-hp pump, assuming the start-up energy is two times the running energy?

Kilowatts = (Number of hp)(0.746 kW/hp)(2)

Kilowatts = (200 hp)(0.746 kW/hp)(2) = 298 kW, round to **300 kW**

125. Calculate the total kilowatts needed to operate a small facility with the following characteristics if everything was running (disregard start-up energy requirements):

Raw water pump	**100 hp**
Five flocculators, 10 hp each	**50 hp**
Filter pumps for backwashing	**100 hp**
Chlorination	**25 hp**
Clear well water pump	**50 hp**
Lighting	**7 hp**
Instrumentation	**3 hp**
First, add the total horsepower:	**335 hp**

The equation is:

$$\textbf{Kilowatts} = \textbf{(Number of hp)(0.746 kW/hp)}$$

$$\text{Kilowatts} = (335 \text{ hp})(0.746 \text{ kW/hp}) = \textbf{250 kW}$$

Water Treatment Grade 2 Test

(answers on p. 119)

1. How many pounds per day of dry alum are needed if the flow rate is 51 mgd and the dosage is 13.8 mg/L. The purity of the alum is 48%.

2. Find the detention time in minutes for a clarifier that has a diameter of 164 ft and a water depth of 11.8 ft if the flow rate is 14.5 mgd.

3. A fluoride dose of 1.20 mg/L is needed for treating a flow of 8,150 gpm. How many pounds per day of sodium fluorosilicate with a commercial purity of 98% and a fluoride ion content of 60.6% will be required? The water being treated contains 0.24 mg/L fluoride.

4. A plant is treating water at 6.95 mgd. If lime is being added at a rate of 135.5 g/min, what are the lime usage in pounds per day and the dosage in milligrams per liter?

5. A polymer solution weighs 152.98 grams per 100 mL. The water treatment plant is using 52 mL/min of the polymer for treating 29.0 ft^3/s. What is the polymer dosage?

6. How many pounds of 62% calcium hypochlorite are required for a 50.0-mg/L dosage in a tank that is 110 ft in diameter and has a water level of 19 ft?

7. Calculate the volume in cubic feet for a 14-in.-diameter pipeline that is 778 ft long.

8. A raw water flow of 31.0 ft^3/s is prechlorinated with 385 lb of chlorine gas. If the flow is changed to 27.0 ft^3/s, what should be the adjustment to the chlorinator?

9. A water treatment plant has eight filters with an average flow rate of 5.84 gpm/ft^2. If the plant flow is 62 ft^3/s, what is the filtration area of each filter?

10. Seventy-three pounds of lime are mixed in a 100-gal tank that contains 85 gal of water. What is the percent of lime in the slurry?

11. Flow through a channel that is 9.2 ft wide is 27 ft^3/s. If the velocity is 2.3 ft/s, what is the depth of the water in the channel?

12. The level in a storage tank drops 5.2 ft in exactly 6 hr. If the tank has a diameter of 90.0 ft and the plant is producing 7.2 mgd, what is the average discharge rate of the treated water discharge pumps in gallons per minute?

Solutions to Water Treatment Grade 2 Test

1. Convert the percent purity to decimal form.

$$48\%/100\% = 0.48$$

The equation is:

$$\textbf{Pounds per day} = \frac{\textbf{(mgd)(Dosage, mg/L)(8.34 lb/gal)}}{\textbf{\% purity}}$$

Substituting:

$$\text{Pounds per day, alum} = \frac{(51\text{ mgd})(13.8\text{ mg/L})(8.34\text{ lb/gal})}{(0.48\text{ purity})}$$

$$= 12{,}228\text{, round to } \textbf{12,000 lb/d, alum}$$

2. First, determine the volume in gallons for the clarifier.

$$\textbf{Volume, gal} = \textbf{(0.785)(Diameter)}^2\textbf{(Depth)(7.48 gal/ft}^3\textbf{)}$$

$$\text{Volume, gal} = (0.785)(164\text{ ft})(164\text{ ft})(11.8\text{ ft})(7.48\text{ gal/ft}^3) = 1{,}863{,}550\text{ gal}$$

Then, convert million gallons per day to gallons per minute (detention time is asked for in minutes).

$$(14.5\text{ mgd})(1{,}000{,}000/1\text{ M})(1\text{ d}/1{,}440\text{ min}) = 10{,}069\text{ gpm}$$

Next, determine the detention time.

The equation is:

$$\textbf{Detention time, min} = \frac{\textbf{Volume, gal}}{\textbf{Flow rate, gpm}}$$

$$\text{Detention time, min} = \frac{1{,}863{,}550\text{ gal}}{10{,}069\text{ gpm}} = \textbf{185 min}$$

3. First, determine how many million gallons per day are being treated.

Million gallons per day = (8,150 gpm)(1,440 min/d)(1 M/1,000,000) = 11.74 mgd

Because natural fluoride (F) is present, subtract the natural from the desired to get the dose required.

F dose required = 1.20 mg/L F – 0.24 mg/L natural F content = 0.96 mg/L F

Write the "pounds" equation with the addition of the percent purity and fluoride content.

$$\textbf{Pounds per day, } \mathbf{Na_2SiF_6} \textbf{ compound} = \frac{\textbf{(mgd)(Dosage, mg/L)(8.34 lb/d)}}{\textbf{(\% purity/100\%)(\% F content/100\%)}}$$

$$\text{Pounds per day, } Na_2SiF_6 = \frac{(11.74 \text{ mgd})(0.96 \text{ mg/L})(8.34 \text{ lb/d})}{(98\% \text{ purity}/100\%)(60.6\% \text{ F content}/100\%)}$$

= 158 lb/d, round to **160 lb/d of Na_2SiF_6**

4. First, find the pounds per day of lime usage.

Pounds per day, lime = (g/min)(1,440 min/d)(1 lb/454 g)

Pounds per day, lime = (135.5 g/min)(1,440 min/d)(1 lb/454 g)

= 429.78 lb/d, round to **430 lb/d of lime**

Then, using the "pounds" equation, calculate the dosage in milligrams per liter by rearranging the formula and solving for dosage.

Next, calculate the dosage.

$$\textbf{Dosage, mg/L} = \frac{\textbf{lb/d}}{\textbf{(mgd)(8.34 lb/gal)}}$$

$$\text{Dosage, mg/L} = \frac{429.78 \text{ lb/d}}{(6.95 \text{ mgd})(8.34 \text{ lb/gal})} = \textbf{7.41 mg/L of lime}$$

5. First, determine the number of grams per milliliter.

152.98 g/100 mL = 1.5298 g/mL

Then, find the number of pounds per gallon for the polymer.

(1.5298 g/mL)(3,785 mL/gal)(1 lb/454 g) = 12.754 lb/gal of polymer

Next, convert cubic feet per second to million gallons per day.

Number of mgd = $(29.0\ ft^3/s)(86{,}400\ s/d)(7.48\ gal/ft^3)(1\ M/1{,}000{,}000) = 18.74$ mgd

Next, find the number of pounds per day of polymer usage.

$$(52\ mL/min)(1{,}440\ min/d)(1\ gal/3{,}785\ mL)(12.754\ lb/gal) = 252.317\ lb/d$$

Then, calculate the dosage.

$$\textbf{Dosage, mg/L} = \frac{\textbf{lb/d}}{\textbf{(mgd)(8.34 lb/gal)}}$$

$$\text{Dosage, mg/L} = \frac{252.317\ lb/d}{(18.74\ mgd)(8.34\ lb/gal)} = \textbf{1.6 mg/L of polymer}$$

6. First, determine the number of million gallons in the tank using the following formula:

$$\textbf{Million gallons} = \textbf{(0.785)(Diameter)}^2\textbf{(Depth)}\frac{\textbf{(7.48 gal)}}{\textbf{ft}^3}\frac{\textbf{(1 M)}}{\textbf{1,000,000}}$$

$$\text{Million gallons} = (0.785)(110\ ft)(110\ ft)(19\ ft)\frac{(7.48\ gal)}{ft^3}\frac{(1\ M)}{1{,}000{,}000} = 1.35\ \text{mil gal}$$

Next, use the "pounds per day" equation, but drop the "day" in this problem.

Pounds of calcium hypochlorite = (mgd)(Dosage)(8.34 lb/gal)

Pounds of calcium hypochlorite = (1.35 mil gal)(50.0 mg/L)(8.34 lb/gal) = 562.95 lb

Because the calcium hypochlorite is not pure (62%), divide the number of pounds by 62% to get the actual number of pounds needed to produce the 50.0-mg/L dose.

$$\frac{562.95\ lb}{62\%/100\%\ Cl_2\ \text{available}}$$

= 907.98 lb, round to **910 lb of calcium hypochlorite needed**

7. First, convert the diameter to feet.

$$\frac{(14\ in.)\ (1\ ft)}{12\ in.} = 1.167\ ft$$

The equation is:

$$\textbf{Volume, ft}^3 = \textbf{(0.785)(Diameter)}^2\textbf{(Length)}$$

$$\text{Volume, } ft^3 = (0.785)(1.167\ ft)(1.167\ ft)(778\ ft)$$

$$\text{Volume, } ft^3 = 831.75\ ft^3\text{, round to } \mathbf{830\ ft^3}$$

8. Solve the equation by setting up a ratio as follows:

$$\frac{\text{lb } Cl_2}{\text{flow}_1} = \frac{x \text{ lb } Cl_2}{\text{flow}_2}$$

Solve for x.

$$x \text{ lb } Cl_2 = \frac{(385 \text{ lb})(27.0 \text{ ft}^3\text{/s})}{31.0 \text{ ft}^3\text{/s}} = \mathbf{335 \textbf{ lb of } Cl_2}$$

9. First, calculate the number of gallons per minute.

$$(62 \text{ ft}^3\text{/s})(7.48 \text{ gal/ft}^3)(60 \text{ s/min}) = 27{,}825.6 \text{ gpm}$$

$$\textbf{Filtration rate} = \frac{\textbf{Flow rate, gpm}}{\textbf{Filter surface area, ft}^2}$$

Rearranging the formula:

$$\text{Filter surface area, ft}^2 = \frac{(\text{Flow rate, gpm})}{(\text{Filtration rate})}$$

$$\text{Filter surface area, ft}^2 = \frac{(27{,}825.6 \text{ gpm})}{(5.84 \text{ gpm/ft}^2)} = 4{,}765 \text{ ft}^2 \text{ (for all eight filters)}$$

$$\text{Filter area for each filter} = 4{,}765 \text{ ft}^2\text{/8 filters}$$

$$= 596 \text{ ft}^2\text{, round to } \textbf{600 ft}^2 \textbf{ for each filter}$$

10.

$$\text{Percent lime} = \frac{(73 \text{ lb})(100\%)}{73 \text{ lb} + (8.34 \text{ lb/gal})(85 \text{ gal})}$$

$$= \frac{(73)(100\%)}{73 \text{ lb} + 708.9 \text{ lb}}$$

$$= \frac{(73 \text{ lb})(100\%)}{781.9 \text{ lb}} = \textbf{9.3\% lime}$$

11. The equation is:

$$\textbf{Q (Flow)} = \textbf{(Area)(Velocity)}$$

$$27 \text{ ft}^3\text{/s} = (9.2 \text{ ft})(x \text{ ft, depth})(2.3 \text{ ft/s})$$

$$x \text{ ft, depth} = \frac{27 \text{ ft}^3\text{/s}}{(9.2 \text{ ft})(2.3 \text{ ft/s})} = \textbf{1.3 ft deep}$$

12. First, find the water production during the 6-hr interval.

Gallons of water treated in 6-hr interval = (7.2 mgd)(1,000,000)(6 hr)/24 hr

= 1,800,000 gal

Next, find the gallons contained in the 5.2-ft drop in water level.

Volume, tank = (0.785)(Diameter)^2(Height)

Volume of 5.2 ft in 90 ft diameter tank = (0.785)(90.0 ft)(90.0 ft)(5.2 ft)(7.48 gal/ft^3)

= 247,320 gal

Production plus the loss in level is the amount the discharge pumps had to send into the distribution system.

Total gallons discharge pumps moved in 6 hr = 1,800,000 gal + 247,320 gal

= 2,047,320 gal

Next, find the number of minutes in 6 hr.

(6 hr)(60 min/hr) = 360 min

Last, divide the number of gallons the discharge pumps moved by the time in minutes.

Discharge pumps, gpm = 2,047,320 gal/360 min

= 5,687gal/min, round to **5,700 gal/min**

CHAPTER 3

WATER TREATMENT

Grade 3

Students preparing for the grade 4 water treatment certification test should also understand these problems.

CONCENTRATION CONVERSIONS

1. Convert 23 lb/mil gal to milligrams per liter.

$$\text{Pounds per million pounds} = \frac{(23 \text{ lb})(1 \text{ gal})}{(1 \text{ mil gal})(8.34 \text{ lb})} = \frac{23 \text{ lb, gal}}{8.34 \text{ lb, mil gal}}$$

The units of pounds and gallons cancel each other out, leaving:

$$\frac{2.76}{\text{M}} = 2.76 \text{ ppm, round to } \mathbf{2.8\ mg/L}$$

***Note*:** 2.8/M is the same as 2.8 ppm, which is the same as 2.8 mg/L.

2. Convert a solution that has 52,600 ppm to percent.

A 1% solution = 10,000 ppm.

$$\frac{52{,}600 \text{ ppm}}{10{,}000 \text{ ppm}/1\%} = \mathbf{5.26\%\ solution}$$

3. Convert 45 lb/mil gal to milligrams per liter.

First, divide 45 lb/mil gal by 8.34 lb/gal.

$$\frac{(45\text{ lb})(1\text{ gal})}{(1\text{ mil gal})(8.34\text{ lb})} = \frac{45\text{ lb, gal}}{8.34\text{ lb, mil gal}}$$

Next, cancel out pounds and gallons.

$$\frac{5.39}{\text{M}} = 5.39\text{ mg/L, round to } \mathbf{5.4\text{ mg/L}}$$

4. Convert 5.7 ppm to pounds per million gallons.

$$5.7\text{ ppm is the same as } \frac{5.7\text{ lb}}{1\text{ M lb}}$$

Now, convert pounds per million pounds to pounds per million gallons by multiplying by 8.34 lb/gal.

$$\frac{(5.7\text{ lb})(8.34\text{ lb/gal})}{1\text{ M lb}} = \mathbf{48\text{ lb/mil gal}}$$

5. Convert a solution that has 138,700 ppm to percent.

A 1% solution = 10,000 ppm.

$$\frac{138{,}700\text{ ppm}}{10{,}000\text{ ppm/1\%}} = \mathbf{13.87\%\text{ solution}}$$

6. Convert 18.5 ppm to pounds per million gallons.

$$18.5 \text{ ppm is the same as } \frac{18.5 \text{ lb}}{1 \text{ M lb}}$$

Convert pounds per million pounds to pounds per million gallons by multiplying by 8.34 lb/gal.

$$\frac{(18.5 \text{ lb})(8.34 \text{ lb/gal})}{1 \text{ M lb}} = \mathbf{154 \text{ lb/mil gal}}$$

7. Convert 54 lb/mil gal to milligrams per liter.

$$\frac{54 \text{ lb}}{1 \text{ mil gal}} = \frac{(54 \text{ lb})(1 \text{ gal})}{(1 \text{ mil gal})(8.34 \text{ lb})}$$

Cancel out like units, pounds and gallons.

$$\frac{6.5}{\text{M}} = \mathbf{6.5 \text{ mg/L}}$$

VOLUME PROBLEMS

8. Find the capacity of a cylindrical tank in cubic feet and in liters if it has a diameter of 14.0 ft and a height of 18.25 ft.

The equation is:

$$\mathbf{Volume = (0.785)(Diameter)^2(Height)}$$

Volume = (0.785)(14.0 ft)(14.0 ft)(18.25 ft) = 2,808 ft^3, round to **2,810 ft^3**

Next, find the number of liters, but use 2,808 ft^3.

(2,808 ft^3)(7.48 gal/ft^3)(3.785 L/gal) = **79,500 L**

9. Find the capacity of a cylindrical tank in cubic feet and in liters if it has a diameter of 13.75 ft and a height of 14.2 ft.

The equation is:

$$\textbf{Volume} = \textbf{(0.785)(Diameter)}^2\textbf{(Height)}$$

$$\text{Volume} = (0.785)(13.75\text{ ft})(13.75\text{ ft})(14.2\text{ ft}) = 2{,}107\text{ ft}^3\text{, round to } \mathbf{2{,}110\ ft^3}$$

Next, find the number of liters.

$$(2{,}107\text{ ft}^3)(7.48\text{ gal/ft}^3)(3.785\text{ L/gal}) = 59{,}653\text{ L, round to } \mathbf{59{,}700\ L}$$

10. Find the capacity of a cylindrical tank in cubic feet and in liters if it has a diameter of 30.0 ft and a height of 24 ft.

The equation is:

$$\textbf{Volume} = \textbf{(0.785)(Diameter)}^2\textbf{(Height)}$$

$$\text{Volume} = (0.785)(30.0\text{ ft})(30.0\text{ ft})(24\text{ ft}) = 16{,}956\text{ ft}^3\text{, round to } \mathbf{17{,}000\ ft^3}$$

Next, find the number of liters.

$$(16{,}956\text{ ft}^3)(7.48\text{ gal/ft}^3)(3.785\text{ L/gal}) = 480{,}055\text{ L, round to } \mathbf{480{,}000\ L}$$

PERCENT DETERMINATIONS

11. A settled sludge test for a solids contact unit indicated 17 mL of sludge had settled after 10 min. What is the settled sludge percent if the sample size was 100 mL?

The equation is:

$$\textbf{Percent settled sludge} = \frac{\textbf{(Settled sludge, mL)(100\%)}}{\textbf{Total sample volume, mL}}$$

$$\text{Percent settled sludge} = \frac{(17\text{ mL})(100\%)}{100\text{ mL}} = \mathbf{17\%}$$

12. What is the percent removal across a settling basin if the influent is 29.1 ntu and the effluent is 0.52 ntu?

The equation is:

$$\textbf{Percent ntu removal} = \frac{\textbf{Influent ntu – Effluent ntu}}{\textbf{Influent}} \times \textbf{100\% or } \frac{\textbf{In – Out}}{\textbf{ntu In}} \times \textbf{100\%}$$

$$\text{Percent ntu removal} = \frac{(29.1 \text{ ntu} - 0.52 \text{ ntu})}{29.1 \text{ ntu}} \times 100\% = \frac{28.58 \text{ ntu}}{29.1 \text{ ntu}} \times 100\% = \mathbf{98\%}$$

PRESSURE PROBLEMS

13. What is the total force in pounds exerted on the bottom of a tank with a diameter of 25 ft if the pressure at the bottom is 125 psig?

First, find the total surface area of the bottom of the tank.

$$\textbf{Area} = \textbf{(0.785)(Diameter)}^2\textbf{(144 in.}^2\textbf{/ft}^2\textbf{)}$$

Substituting:

$$\text{Area} = (0.785)(25 \text{ ft})(25 \text{ ft})(144 \text{ in.}^2/\text{ft}^2)$$

$$\text{Area} = 70{,}650 \text{ in.}^2$$

Then, calculate the force in pounds.

$$\textbf{Total force, pounds} = \textbf{(Area, in.}^2\textbf{)(psig)}$$

Total force, pounds = $(70{,}650 \text{ in.}^2)(125 \text{ psig})$ = 8,831,250 lb, round to **8,800,000 lb**

14. A rectangular tank measures 8 ft by 12 ft. Water in the tank is 12 ft in depth. What is the pressure in pounds per square inch on the bottom of the tank?

Pressure, in psi on bottom of tank = (0.433 psi/ft)(12 ft) = **5.2 psi**

15. What is the pressure in pounds per square inch at the bottom of a polymer storage tank if the level of the polymer is 10.3 ft? The density of the polymer is 9.89 lb/gal.

The equation is:

$$\text{psi} = \frac{\text{Depth, ft}}{2.31\ \text{ft/psi}}$$

$$\text{psi} = \frac{10.3\ \text{ft}}{2.31\ \text{ft/psi}} = 4.46\ \text{psi}$$

Next, find the specific gravity of the polymer.

$$\text{Specific gravity} = \frac{9.89\ \text{lb/gal for polymer}}{8.34\ \text{lb/gal for water}} = 1.19$$

Then, multiply the psi by the specific gravity to determine the psi.

$$\text{psi} = (4.46\ \text{psi})(1.19) = \mathbf{5.31\ psi}$$

16. A rectangular tank measures 8 ft by 15 ft. Water in the tank is 7.25 ft in depth. What is the pressure in pounds per square inch on the bottom of the tank?

The following applies: 1 psi = 2.31 ft.

$$\frac{(\text{Number of ft})(1\ \text{psi})}{(2.31\ \text{ft})}$$

Substituting:

$$\frac{(7.25\ \text{ft})(1\ \text{psi})}{(2.31\ \text{ft})} = 3.14\ \text{psi}$$

17. What is the pressure in pounds per square inch at the bottom of a polymer storage tank if the level of the alum is 12.4 ft? The density of the alum is 10.27 lb/gal.

The equation is:

$$\text{psi} = \frac{\text{Depth, ft}}{2.31 \text{ ft/psi}}$$

$$\text{psi} = \frac{12.4 \text{ ft}}{2.31 \text{ ft/psi}} = 5.37 \text{ psi}$$

Next, find the specific gravity of the alum.

$$\text{Specific gravity} = \frac{10.27 \text{ lb/gal for alum}}{8.34 \text{ lb/gal for water}} = 1.23$$

Then, multiply the psi by the specific gravity to determine the psi.

psi = (5.37 psi)(1.23) = **6.61 psi**

18. A rectangular tank measures 8.5 ft by 15 ft. Water in the tank is 7 ft deep. What is the pressure in pounds per square inch at the bottom of the tank?

Pressure, at psi at bottom of tank = (0.433 psi/ft)(7 ft) = **3 psi**

DETENTION TIME PROBLEMS

19. Calculate the theoretical detention time in hours for a sedimentation (sed) basin that is 690 ft long, 48 ft wide, and has a water depth of 8.5 ft if the flow is 12.2 mgd.

First, determine the number of gallons in the sed basin.

The equation is:

$$\textbf{Volume, gal} = \textbf{(Length)(Width)(Depth)(7.48 gal/ft}^3\textbf{)}$$

$$\text{Volume, gal} = (690 \text{ ft})(48 \text{ ft})(8.5 \text{ ft})(7.48 \text{ gal/ft}^3) = 2{,}105{,}770 \text{ gal}$$

Next, convert million gallons per day to gallons per hour.

$$(12.2 \text{ mgd})(1 \text{ d}/24 \text{ hr})(1{,}000{,}000/1 \text{ M}) = 508{,}333 \text{ gph}$$

Write the equation with units asked for in the question.

$$\textbf{Detention time, hr} = \frac{\textbf{Volume, gal}}{\textbf{Flow rate, gph}}$$

Substituting:

$$\text{Detention time, hr} = \frac{2{,}105{,}770 \text{ gal}}{508{,}333 \text{ gph}} = 4.14 \text{ hr, round to } \mathbf{4.1 \text{ hr}}$$

20. Calculate the theoretical detention time in hours for a plant's flocculation and sedimentation (sed) basin if the flow is 8.9 mgd. Five flocculation basins are 40 ft by 15 ft with a water depth of 12 ft and one sed basin is 400.0 ft long, 60.0 ft wide, and has a water depth of 9.0 ft.

First, determine the number of gallons in the five flocculation basins and the sed basin.

The equation is:

$$\textbf{Volume, gal} = \textbf{(Length)(Width)(Depth)(7.48 gal/ft}^3\textbf{)(5 basins)}$$

$$\text{Volume, gal in floc basins} = (40 \text{ ft})(15 \text{ ft})(12 \text{ ft})(7.48 \text{ gal/ft}^3)(5 \text{ basins}) = 269{,}280 \text{ gal}$$

$$\text{Volume, gal in sed basin} = (400.0 \text{ ft})(60.0 \text{ ft})(9.0 \text{ ft})(7.48 \text{ gal/ft}^3) = \underline{1{,}615{,}680 \text{ gal}}$$

$$1{,}884{,}960 \text{ gal}$$

Next, convert million gallons per day to gallons per hour.

$$(8.9 \text{ mgd})(1 \text{ d}/24 \text{ hr})(1{,}000{,}000/1 \text{ M}) = 370{,}833 \text{ gph}$$

Write the equation with units asked for in the question.

$$\textbf{Detention time, hr} = \frac{\textbf{Volume, gal}}{\textbf{Flow rate, gph}}$$

$$\text{Detention time, hr} = \frac{1{,}884{,}960 \text{ gal}}{370{,}833 \text{ gph}} = \textbf{5.1 hr}$$

21. Find the theoretical detention time in minutes for a clarifier that has a diameter of 120 ft and a water depth of 14 ft if the flow rate is 1.84 mgd.

First, determine the volume in gallons for the clarifier.

$$\textbf{Volume, gal} = \textbf{(0.785)(Diameter)}^2\textbf{(Depth)(7.48 gal/ft}^3\textbf{)}$$

$$\text{Volume, gal} = (0.785)(120 \text{ ft})(120 \text{ ft})(14 \text{ ft})(7.48 \text{ gal/ft}^3) = 1{,}183{,}755 \text{ gal}$$

Then, convert million gallons per day to gallons per minute, because detention time is asked for in minutes.

$$(1.84 \text{ mgd})(1{,}000{,}000/1 \text{ M})(1 \text{ d}/1{,}440 \text{ min}) = 1{,}278 \text{ gpm}$$

The equation is:

$$\textbf{Detention time, min} = \frac{\textbf{Volume, gal}}{\textbf{Flow rate, gpm}}$$

$$\text{Detention time, min} = \frac{1{,}183{,}755 \text{ gal}}{1{,}278 \text{ gpm}} = 926.26 \text{ min, round to } \textbf{930 min}$$

22. Find the theoretical detention time in hours for a clarifier that has a diameter of 100.0 ft and a water depth of 12.5 ft if the flow rate is 1.72 mgd.

First, determine the volume in gallons for the clarifier.

$$\textbf{Volume, gal} = \mathbf{(0.785)(Diameter)^2(Depth)(7.48\ gal/ft^3)}$$

$$\text{Volume, gal} = (0.785)(100.0\text{ ft})(100.0\text{ ft})(12.5\text{ ft})(7.48\text{ gal/ft}^3)$$

$$\text{Volume, gal} = 733{,}975\text{ gal}$$

Then, convert million gallons per day to gallons per minute, because detention time is asked for in minutes.

$$(1.72\text{ mgd})(1{,}000{,}000/1\text{ M})(1\text{ d}/24\text{ hr}) = 71{,}667\text{ gph}$$

The equation is:

$$\textbf{Detention time, hr} = \frac{\textbf{Volume, gal}}{\textbf{Flow rate, gph}}$$

$$\text{Detention time, hr} = \frac{733{,}975\text{ gal}}{71{,}667\text{ gph}} = \mathbf{10.2\ hr}$$

23. Calculate the detention time in hours for four flocculation basins each 50 ft by 12 ft with a water depth of 11 ft and a sedimentation basin that is 380 ft long, 70 ft wide, and has a water depth of 10 ft. The flow is 10.8 mgd.

First, determine the number of gallons in the four flocculation basins and the sedimentation basin.

The equation is:

$$\textbf{Volume, gal} = \mathbf{(Length)(Width)(Depth)(7.48\ gal/ft^3)(5\ basins)}$$

$$\text{Volume, gal in floc basins} = (50\text{ ft})(12\text{ ft})(11\text{ ft})(7.48\text{ gal/ft}^3)(4\text{ basins}) = 197{,}472\text{ gal}$$

$$\text{Volume, gal in sed basin} = (380\text{ ft})(70\text{ ft})(10\text{ ft})(7.48\text{ gal/ft}^3) = \underline{1{,}989{,}680\text{ gal}}$$

$$2{,}187{,}152\text{ gal}$$

Next, convert million gallons per day to gallons per hour.

$$(10.8\text{ mgd})(1\text{ d}/24\text{ hr})(1{,}000{,}000/1\text{M}) = 450{,}000\text{ gph}$$

Write the equation with units asked for in the question.

$$\textbf{Detention time, hr} = \frac{\textbf{Volume, gal}}{\textbf{Flow rate, gph}}$$

$$\text{Detention time, hr} = \frac{2{,}187{,}152 \text{ gal}}{450{,}000 \text{ gph}} = 4.86 \text{ hr, round to } \mathbf{5\ hr}$$

24. Find the detention time in hours for a treatment plant that includes the following: A sedimentation (sed) basin 750 ft long, 75.0 ft wide, and with a water depth of 11.0 ft; eight filters each 36.0 ft long, 24.0 ft wide, and with an average water depth of 9.0 ft; flow is 7.82 mgd.

First, determine the volume in gallons for the sed basin.

$$\text{Volume, gal} = (750 \text{ ft})(75.0 \text{ ft})(11.0 \text{ ft})(7.48 \text{ gal/ft}^3) = 4{,}628{,}250 \text{ gal}$$

Then, find the volume of the filters.

$$\text{Volume, gal} = (36.0 \text{ ft})(24.0 \text{ ft})(9.0 \text{ ft})(8 \text{ filters})(7.48 \text{ gal/ft}^3) = 465{,}316 \text{ gal}$$

The total volume would then be the sum of the filters and sed basin, but first convert million gallons per day to gallons per hour.

$$(7.82 \text{ mgd})(1{,}000{,}000/1 \text{ M})(1 \text{ d}/24 \text{ hr}) = 325{,}833 \text{ gph}$$

Then, find the total volume.

$$\text{Total volume, gal} = 4{,}628{,}250 + 465{,}316 = 5{,}093{,}566 \text{ gal}$$

The equation is:

$$\textbf{Detention time, hr} = \frac{\textbf{Volume, gal}}{\textbf{Flow rate, gph}}$$

$$\text{Detention time, hr} = \frac{5{,}093{,}566 \text{ gal}}{325{,}833 \text{ gph}} = \mathbf{15.6\ hr}$$

25. Find the detention time in minutes for a clarifier that has a diameter of 160 ft and a water depth of 16 ft if the flow rate is 3.9 mgd.

First, determine the volume in gallons for the clarifier.

$$\textbf{Volume, gal} = \mathbf{(0.785)(Diameter)^2(Depth)(7.48\ gal/ft^3)}$$

$$\text{Volume, gal} = (0.785)(160\text{ ft})(160\text{ ft})(16\text{ ft})(7.48\text{ gal/ft}^3) = 2{,}405{,}089\text{ gal}$$

Then, convert million gallons per day to gallons per minute, because detention time is asked for in minutes.

$$(3.9\text{ mgd})(1{,}000{,}000/1\text{ M})(1\text{ d}/1{,}440\text{ min}) = 2{,}708\text{ gpm}$$

The equation is:

$$\textbf{Detention time, min} = \frac{\textbf{Volume, gal}}{\textbf{Flow rate, gpm}}$$

$$\text{Detention time, min} = \frac{2{,}405{,}089\text{ gal}}{2{,}708\text{ gpm}} = 888\text{ min, round to }\mathbf{890\ min}$$

26. Find the detention time in hours for a treatment plant that includes the following: A sedimentation (sed) basin 272 ft long, 79.0 ft wide, and with an average water depth of 11.5 ft; 12 filters each 40.0 ft long, 32.0 ft wide, and with an average water depth of 9.00 ft; flow is 16.50 mgd.

First, determine the volume in gallons for the sedimentation basin.

$$\text{Volume, gal} = (272\text{ ft})(79.0\text{ ft})(11.5\text{ ft})(7.48\text{ gal/ft}^3) = 1{,}848{,}398\text{ gal}$$

Then, find the volume of the filters.

$$\text{Volume, gal} = (40.0\text{ ft})(32.0\text{ ft})(9.00\text{ ft})(12.0\text{ filters})(7.48\text{ gal/ft}^3) = 1{,}034{,}035\text{ gal}$$

The total volume would then be the sum of the filters and sedimentation basin.

$$\text{Total volume, gal} = 1{,}848{,}398\text{ gal} + 1{,}034{,}035\text{ gal} = 2{,}882{,}433\text{ gal}$$

Next, convert million gallons per day to gallons per hour.

$$(16.50\text{ mgd})(1{,}000{,}000/1\text{ M})(1\text{ d}/24\text{ hr}) = 687{,}500\text{ gph}$$

The equation is:

$$\textbf{Detention time, hr} = \frac{\textbf{Volume, gal}}{\textbf{Flow rate, gph}}$$

$$\text{Detention time, hr} = \frac{2{,}882{,}433\ \text{gal}}{687{,}500\ \text{gph}} = \textbf{4.19 hr}$$

DOSAGE PROBLEMS

27. A water treatment plant is treating 3.5 mgd with 44 lb/d of chlorine. What is the chlorine dosage in milligrams per liter?

The "pounds per day" equation is:

$$\textbf{Number of lb/d of chlorine} = \textbf{(mgd)(Dosage, mg/L)(8.34 lb/gal)}$$

Substituting:

$$44\ \text{lb/d} = (3.5\ \text{mgd})(\text{Dosage, mg/L})(8.34\ \text{lb/gal})$$

Solve for dosage.

$$\text{Dosage} = \frac{44\ \text{lb/d}}{(3.5\ \text{mgd})(8.34\ \text{lb/gal})} = \textbf{1.5 mg/L, chlorine}$$

28. The dosage for chlorine at a treatment plant is 650 lb/d for a flow of 21 ft³/s. If the flow is adjusted to 17 ft³/s, what would the theoretical chlorine dosage be in pounds per day if everything else remains the same?

$$\frac{650\ \text{lb/d}}{21\ \text{ft}^3/\text{s}} = \frac{x\text{, lb/d}}{17\ \text{ft}^3/\text{s}}$$

Solve for x.

$$x\text{, mL/min} = \frac{(650\ \text{lb/d})(17\ \text{ft}^3/\text{s})}{21\ \text{ft}^3/\text{s}} = 526\ \text{lb/d, round to } \textbf{530 lb/d of chlorine}$$

29. A water treatment plant is treating 18.5 mgd with 304 lb/d of chlorine. What is the chlorine dosage in milligrams per liter?

The equation is:

$$\textbf{Number of lb/d of chlorine} = \textbf{(mgd)(Dosage, mg/L)(8.34 lb/gal)}$$

$$304 \text{ lb/d} = (18.5 \text{ mgd})(\text{Dosage, mg/L})(8.34 \text{ lb/gal})$$

$$\text{Dosage, mg/L} = \frac{304 \text{ lb/d}}{(18.5 \text{ mgd})(8.34 \text{ lb/gal})} = \textbf{1.97 mg/L, chlorine}$$

30. What should the chemical feeder setting be in pounds per day and pounds per minute if 4.52 mgd is treated with 12.3 mg/L of soda ash?

The equation is:

$$\textbf{Number of lb/d of soda ash} = \textbf{(mgd)(Dosage, mg/L)(8.34 lb/gal)}$$

$$\text{Pounds per day of soda ash} = (4.52 \text{ mgd})(12.3 \text{ mg/L})(8.34 \text{ lb/gal})$$

$$\text{Pounds per day of soda ash} = 463.67 \text{ lb/d, round to } \textbf{464 lb/d of soda ash}$$

Then, convert to pounds per minute, but use 463.67 lb/d.

$$\frac{463.67 \text{ lb/d}}{1{,}440 \text{ min/d}} = \textbf{0.322 lb/min of soda ash}$$

31. What should the chemical feeder setting be in milliliters per minute for a polymer solution if the desired dosage is 2.00 mg/L and the treatment plant is treating 20.6 mgd? The specific gravity of the polymer is 1.26.

Find the number of pounds per day of polymer required by using the "pounds" equation.

$$\textbf{Pounds per day, polymer} = \textbf{(mgd)(Dosage, mg/L)(8.34 lb/gal)}$$

$$\text{Pounds per day, polymer} = (20.6 \text{ mgd})(2.00 \text{ mg/L})(8.34 \text{ lb/gal}) = 343.6 \text{ lb/d}$$

Next, determine the pounds per gallon of the polymer solution.

$$\text{Pounds per gallon} = (\text{Specific gravity})(8.34 \text{ lb/gal}) = (1.26)(8.34 \text{ lb/gal}) = 10.5 \text{ lb/gal}$$

Convert the number of pounds per day to number of gallons per day.

$$\text{Gallons per day, polymer} = \frac{343.6 \text{ lb/d}}{10.5 \text{ lb/gal}} = 32.7 \text{ gal/d}$$

Then, convert gallons per day to milliliters per minute.

$$\text{Milliliters per minute, polymer} = \frac{(32.7 \text{ gal/d})(3{,}785 \text{ mL/gal})}{1{,}440 \text{ min/d}} = \mathbf{86.0\ mL/min\ of\ polymer}$$

32. How many pounds of 64% calcium hypochlorite are required for a 2.5-mg/L dosage for a tank that is 80.0 ft in diameter and has a water level of 16.0 ft?

First, determine the number of million gallons in the tank with the following formula:

$$\text{Million gallons} = (0.785)(\text{Diameter})^2(7.48 \text{ gal/ft}^3)(1 \text{ M}/1{,}000{,}000)$$

$$\text{Million gallons} = (0.785)(80.0 \text{ ft})(80.0 \text{ ft})(16.0 \text{ ft})(7.48 \text{ gal/ft}^3)(1 \text{ M}/1{,}000{,}000)$$

$$= 0.601 \text{ mil gal}$$

Next, use the "pounds per day" equation, but drop the "day" in this particular problem.

Pounds, calcium hypochlorite = (mgd)(Dosage, mg/L)(8.34 lb/gal)

$$\text{Pounds, calcium hypochlorite} = (0.601 \text{ mil gal})(2.5 \text{ mg/L})(8.34 \text{ lb/gal}) = 12.53 \text{ lb}$$

The calcium hypochlorite is only 64% pure, so divide the above result by 64%.

$$\frac{12.53 \text{ lb required}}{64\% \text{ available } Cl_2/100\%} = 19.578 \text{ lb, round to } \mathbf{20\ lb\ calcium\ hypochlorite\ needed}$$

33. A conventional treatment plant processes 6,450 gpm. If the lime dosage is 425 g/min, how many pounds of lime will the plant use in one month (30 days)? What is the dosage in milligrams per liter?

First, convert grams per minute of lime to pounds per day.

$$(425 \text{ g/min})(1 \text{ lb}/454 \text{ g})(1{,}440 \text{ min/d}) = 1{,}348 \text{ lb/d}$$

Next, determine the number of pounds used in 30 days.

$$(1{,}348 \text{ lb/d})(30 \text{ d/month}) = \mathbf{40{,}440 \text{ lb/month of lime}}$$

Next, convert gallons per minute to million gallons per day.

$$(6{,}450 \text{ gpm})(1{,}440 \text{ min/d})(1 \text{ M}/1{,}000{,}000) = 9.288 \text{ mgd}$$

Then, find the dosage by using the "pounds" equation and rearranging to solve for dosage.

The equation is:

$$\mathbf{Number\ lb/d = (mgd)(Dosage,\ mg/L)(8.34\ lb/gal)}$$

$$\text{Dosage, mg/L} = \frac{\text{lb/d of lime}}{(\text{mgd})(8.34 \text{ lb/gal})}$$

$$\text{Lime dosage, mg/L} = \frac{1{,}348 \text{ lb/d}}{(9.288 \text{ mgd})(8.34 \text{ lb/d})} = \mathbf{17.4\ mg/L\ of\ lime}$$

34. A treatment plant removes 389 lb/d of solids. If the concentration of solids in the raw water averages 4.6 mg/L with a removal efficiency of 96%, what is the production of the plant in million gallons per day?

The equation is:

$$\mathbf{Number\ of\ lb/d = (mgd)(Dosage,\ mg/L)(8.34\ lb/gal)(\%\ removal\ efficiency)}$$

Rearrange to solve for millions gallons per day and substitute known values.

$$= \frac{(389 \text{ lb/d})(96\%/100\%)}{(4.6 \text{ mg/L})(8.34 \text{ lb/gal})}$$

Million gallons per day = 9.73 mgd, round to **9.7 mgd**

35. Water is being delivered to a plant at a pumping rate of 650 gpm. If alum is being added at 14 mg/L, how many pounds of alum are used each day?

First, convert the pumping rate of 650 gpm to million gallons per day.

(650 gpm)(1,440 min/d)(1 M/1,000,000) = 0.936 mgd

The equation is:

Number of lb/d = (mgd)(Dosage, mg/L)(8.34 lb/gal)

Pounds per day, alum = (0.936 mgd)(14 mg/L)(8.34 lb/gal)

Pounds per day, alum = 109.29 lb/d, round to **110 lb/d, alum**

36. How many pounds per day of 60% calcium hypochlorite are required to maintain a 4.0-mg/L dosage for a 3,250-gpm treatment plant?

First, determine the number of million gallons per day.

$$\text{Million gallons per day} = (3{,}250\text{ gpm})(1{,}440\text{ min/d})\frac{(1\text{ M})}{1{,}000{,}000} = 4.68\text{ mgd}$$

Pounds per day, calcium hypochlorite = (mgd)(Dosage, mg/L)(8.34 lb/gal)

Pounds per day, calcium hypochlorite = (4.68 mgd)(4.0 mg/L)(8.34 lb/gal) = 156 lb/d required

$$\frac{156\text{ lb/d required}}{60\%\text{ available } Cl_2/100\%} = \textbf{260 lb/d calcium hypochlorite needed}$$

37. How many pounds per day of 61% calcium hypochlorite are required to maintain a 2.0-mg/L dosage for a treatment plant producing 8,625 gpm?

First, determine the number of million gallons per day.

$$\text{Million gallons per day} = (8{,}625 \text{ gpm})(1{,}440 \text{ min/d}) \frac{(1 \text{ M})}{1{,}000{,}000} = 12.42 \text{ mgd}$$

Pounds per day, calcium hypochlorite = (mgd)(Dosage, mg/L)(8.34 lb/gal)

$$\text{Pounds per day, calcium hypochlorite} = (12.42 \text{ mgd})(2.0 \text{ mg/L})(8.34 \text{ lb/gal})$$

$$= 207.17 \text{ lb/d required}$$

$$\frac{207.17 \text{ lb/d required}}{61\% \text{ available } Cl_2/100\%} = 339.6, \text{ round to } \mathbf{340 \text{ lb/d calcium hypochlorite needed}}$$

38. How many pounds of 64% calcium hypochlorite are required for a 50.0-mg/L dosage for a tank that is 60.0 ft in diameter and has a water level of 24 ft?

First, determine the number of million gallons in the tank using the following formula:

$$\mathbf{Million\ gallons = (0.785)(Diameter)^2(Depth)(7.48\ gal) \frac{(1\ M)}{1{,}000{,}000}}$$

$$\text{Million gallons} = (0.785)(60.0 \text{ ft})(60.0 \text{ ft})(24 \text{ ft})(7.48 \text{ gal}) \frac{(1 \text{ M})}{1{,}000{,}000} = 0.507 \text{ mil gal}$$

Next, use the "pounds per day" equation, but drop the "day" in this case.

Pounds, calcium hypochlorite = (mgd)(Dosage, mg/L)(8.34 lb/gal)

$$\text{Pounds, calcium hypochlorite} = (0.507 \text{ mil gal})(50.0 \text{ mg/L})(8.34 \text{ lb/gal}) = 211.42 \text{ lb}$$

The calcium hypochlorite is only 64% pure, so divide by 64%.

$$\frac{211.42 \text{ lb/d required}}{64\% \text{ available } Cl_2/100\%}$$

$$= 330.34 \text{ lb, round to } \mathbf{330 \text{ lb of calcium hypochlorite needed}}$$

39. A fluoride dose of 1.20 mg/L is needed to treat a flow of 2,850 gpm. How many pounds per day of sodium fluorosilicate (Na_2SiF_6) with a commercial purity of 98% and a fluoride ion content of 60.6% will be required? The water being treated contains 0.12 mg/L fluoride.

First, determine how many million gallons per day are being treated.

Million gallons per day = (2,850 gpm)(1,440 min/d)(1 M/1,000,000) = 4.10 mgd

Because natural fluoride (F) is present, subtract the amount of natural fluoride from the desired amount to get the dose required.

F dose required = 1.20 mg/L F – 0.12 mg/L natural F content = 1.08 mg/L F

Write the "pounds" equation with the addition of the percent purity and fluoride (F) content.

$$\textbf{Pounds per day, } \mathbf{Na_2SiF_6} \textbf{ compound} = \frac{\textbf{(mgd)(Dosage, mg/L)(8.34 lb/d)}}{\textbf{(\% purity/100\%)(\% F content/100\%)}}$$

$$\text{Pounds per day, } Na_2SiF_6 \text{ compound} = \frac{(4.10 \text{ mgd})(1.08 \text{ mg/L})(8.34 \text{ lb/d})}{(98\% \text{ purity}/100\%)(60.6\% \text{ F content}/100\%)}$$

= 62.18 lb/d, round to **62 lb/d of Na_2SiF_6**

40. A fluoride dose of 1.0 mg/L is used to treat a flow of 7 mgd. How many pounds per day of sodium silicofluoride with a commercial purity of 98% and a fluoride ion content of 60.6% are needed? The water being treated contains 0.15 mg/L fluoride.

Write the "pounds" equation with the addition of the percent purity and fluoride (F) content.

$$\textbf{Pounds per day, } \mathbf{Na_2SiF_6} \textbf{ compound} = \frac{\textbf{(mgd)(Dosage, mg/L)(8.34 lb/d)}}{\textbf{(\% purity/100\%)(\% F content/100\%)}}$$

$$\text{Pounds per day, } Na_2SiF_6 \text{ compound} = \frac{(7 \text{ mgd})(1.0 \text{ mg/L})(8.34 \text{ lb/d})}{(98\% \text{ purity}/100\%)(60.6\% \text{ F content}/100\%)}$$

= 98.3 lb/d, round to **100 lb/d of Na_2SiF_6**

41. How many pounds of 64% calcium hypochlorite are required for a 2.0-mg/L dosage for a tank that is 100.0 ft in diameter and has a water level of 28 ft?

First, determine the number of million gallons in the tank using the following formula:

$$\text{Million gallons} = (0.785)(\text{Diameter})^2(\text{Depth})\frac{(7.48\text{ gal})}{\text{ft}^3}\frac{(1\text{ M})}{1{,}000{,}000}$$

$$\text{Million gallons} = (0.785)(100.0\text{ ft})(100.0\text{ ft})(28\text{ ft})\frac{(7.48\text{ gal})}{\text{ft}^3}\frac{(1\text{ M})}{1{,}000{,}000} = 1.64\text{ mil gal}$$

Next, use the "pounds per day" equation.

$$\text{Pounds per day, chlorine} = (\text{mgd})(\text{Dosage, mg/L})(8.34\text{ lb/gal})$$

***Note*:** Drop the D for day in this case.

$$\text{Pounds of chlorine} = (1.64\text{ mil gal})(2.0\text{ mg/L})(8.34\text{ lb/gal}) = 27.36\text{ lb required}$$

$$\frac{27.36\text{ lb required}}{64\%\text{ Available }Cl_2/100\%} = 42.75\text{ lb, round to } \mathbf{43\text{ lb calcium hypochlorite needed}}$$

42. A well that is 220 ft deep and has a diameter of 14 in. requires disinfection. Depth to water from the top of the casing is 83 ft. If the desired dose is 50.0 mg/L, how many pounds of calcium hypochlorite (65% available chlorine) are required?

First, find the length (in feet) of the water-filled casing.

Length of water-filled casing = Depth of well – Depth of water to top of casing

$$\text{Length of water-filled casing} = 220\text{ ft} - 83\text{ ft} = 137\text{ ft}$$

Then, convert the diameter from inches to feet.

$$\text{Diameter, ft} = \frac{14\text{ in.}}{12\text{ in./ft}} = 1.167\text{ ft}$$

Next, determine the volume of water in the well casing (in gallons) using the following formula:

$$\mathbf{Volume, gal = (0.785)(Diameter)^2(Length)(7.48\ gal/ft^3)}$$

$$\text{Volume, gal} = (0.785)(1.167\text{ ft})(1.167\text{ ft})(137\text{ ft})(7.48\text{ gal/ft}^3) = 1{,}096\text{ gal}$$

Next, determine the number of million gallons.

$$\text{Million gallons} = (1{,}096 \text{ gal})(1 \text{ M}/1{,}000{,}000) = 0.001096 \text{ mil gal}$$

Last, using the "pounds" formula, calculate the number of pounds of calcium hypochlorite.

$$\text{Pounds, calcium hypochlorite} = \frac{(0.001096 \text{ mil gal})(50.0 \text{ mg/L})(8.34 \text{ lb/gal})}{(65\% \text{ Available chlorine}/100\%)}$$

$$\text{Pounds, calcium hypochlorite} = \mathbf{0.7\ lb}$$

43. A well is to be disinfected with 60.0% calcium hypochlorite. The well is 365 ft in depth and 1.5 ft in diameter. Depth to water from the top of the casing is 208 ft. If the desired dose is 50.0 mg/L, how many pounds of calcium hypochlorite are required?

First, find the length (in feet) of the water-filled casing.

$$\textbf{Length of water-filled casing} = \textbf{Depth of well} - \textbf{Depth of water to top of casing}$$

$$\text{Length of water-filled casing} = 365 \text{ ft} - 208 \text{ ft} = 157 \text{ ft}$$

Then, determine the volume in gallons of water in the well casing using the following formula:

$$\textbf{Volume, gal} = \mathbf{(0.785)(Diameter)^2(Length)(7.48\ gal/ft^3)}$$

$$\text{Volume, gal} = (0.785)(1.5 \text{ ft})(1.5 \text{ ft})(157 \text{ ft})(7.48 \text{ gal/ft}^3) = 2{,}074 \text{ gal}$$

Next, determine the number of million gallons.

$$\text{Million gallons} = (2{,}074 \text{ gal})(1 \text{ M}/1{,}000{,}000) = 0.002074 \text{ mil gal}$$

Last, using the "pounds" equation, calculate the number of pounds of calcium hypochlorite.

$$\text{Pounds, calcium hypochlorite} = \frac{(0.002074 \text{ mil gal})(50.0 \text{ mg/L})(8.34 \text{ lb/gal})}{(60.0\% \text{ Available chlorine}/100\%)}$$

$$\text{Pounds, calcium hypochlorite} = 1.44 \text{ lb, round to } \mathbf{1.4\ lb}$$

44. A polymer solution weighs 1.18 g/mL. The water treatment plant is using 14 mL/min of the polymer for treating 7.5 ft^3/s. What is the polymer dosage?

First, find number of pounds per gallon for the polymer.

$$(1.18 \text{ g/mL})(3{,}785 \text{ mL/gal})(1 \text{ lb}/454 \text{ g}) = 9.838 \text{ lb/gal}$$

Next, convert cubic feet per second to million gallons per day.

$$(7.5 \text{ ft}^3\text{/s})(86{,}400 \text{ s/d})(7.48 \text{ gal/ft}^3)(1 \text{ M}/1{,}000{,}000) = 4.847 \text{ mgd}$$

Then, find the number of pounds per day of polymer usage.

$$(14 \text{ mL/min})(1{,}440 \text{ min/d})(1 \text{ gal}/3{,}785 \text{ mL})(9.838 \text{ lb/gal}) = 52.4 \text{ lb/d}$$

Now, determine the dosage.

$$\textbf{Dosage, mg/L} = \frac{\textbf{lb/d}}{\textbf{(mgd)(8.34 lb/gal)}}$$

$$\text{Dosage, mg/L} = \frac{52.4 \text{ lb/d}}{(4.847 \text{ mgd})(8.34 \text{ lb/gal})}$$

= 1.296 mg/L, round to **1.3 mg/L of polymer**

45. A plant is treating water at 12.4 mgd. If lime is being added at a rate of 220.7 g/min, what are the lime usage in pounds per day and the dosage in milligrams per liter?

First, find the pounds per day of lime usage.

$$\textbf{Pounds per day, lime} = \textbf{(g/min)(1,440 min/d)(1 lb/454 g)} = \textbf{lb/d}$$

Pounds per day, lime = (220.7 g/min)(1,440 min/d)(1 lb/454 g) = **700 lb/d of lime**

Then, using the "pounds" equation, calculate the dosage in milligrams per liter by rearranging the formula and solving for dosage.

$$\textbf{Dosage, mg/L} = \frac{\textbf{lb/d}}{\textbf{(mgd)(8.34 lb/gal)}}$$

$$\text{Dosage, mg/L} = \frac{700 \text{ lb/d}}{(12.4 \text{ mgd})(8.34 \text{ lb/gal})} = 6.769 \text{ mg/L, round to } \textbf{6.77 mg/L of lime}$$

46. A treatment plant is adding 408.33 g/min of soda ash to its treated water. If the plant is producing water at 32 mgd, what are the soda ash usage in pounds per day and dosage in milligrams per liter?

First, find the pounds per day of soda ash usage.

Pounds per day, soda ash = (g/min)(1,440 min/d)(1 lb/454 g)

Pounds per day, soda ash = (408.33 g/min)(1,440 min/d)(1 lb/454 g)

= 1,295.14 lb/d, round to **1,300 lb/d of soda ash**

Then, using the "pounds" equation, calculate the dosage in milligrams per liter.

$$\text{Dosage, mg/L} = \frac{1{,}295.14 \text{ lb/d}}{(32 \text{ mgd})(8.34 \text{ lb/gal})} = 4.85 \text{ mg/L, round to } \mathbf{4.9 \text{ mg/L of soda ash}}$$

47. A polymer solution weighs 13.25 g per 10.0 mL. The water treatment plant is using 22 mL/min of the polymer for treating 14 ft^3/s. What is the polymer dosage?

First, determine the number of grams per milliliter.

13.25 g/10.0 mL = 1.325 g/mL

Then, find the number of pounds per gallon for the polymer.

(1.325 g/mL)(3,785 mL/gal)(1 lb/454 g) = 11.047 lb/gal

Next, convert cubic feet per second to million gallons per day.

(14 ft^3/s)(86,400 s/d)(7.48 gal/ft^3)(1 M/1,000,000) = 9.048 mgd

Then, find the number of pounds per day of polymer usage.

(22 mL/min)(1,440 min/d)(1 gal/3,785 mL)(11.047 lb/gal) = 92.462 lb/d

Now, find the dosage.

$$\mathbf{Dosage, mg/L} = \frac{\mathbf{lb/d}}{\mathbf{(mgd)(8.34\ lb/gal)}}$$

$$\text{Dosage, mg/L} = \frac{92.462 \text{ lb/d}}{(9.048 \text{ mgd})(8.34 \text{ lb/gal})}$$

= 1.225 mg/L, round to **1.2 mg/L of polymer**

48. A conventional treatment plant processes 2,925 gpm on the average for a one-month period. If the lime dosage is 133 g/min, how many pounds of lime will the plant use in one month (exactly 30 days)? What is the average dosage in milligrams per liter?

First, convert gallons per minute to million gallons per day.

$$(2{,}925 \text{ gpm})(1{,}440 \text{ min/d})(1 \text{ M}/1{,}000{,}000) = 4.212 \text{ mgd}$$

Next, convert grams per minute of lime to pounds per day.

$$(133 \text{ g/min})(1 \text{ lb}/454 \text{ g})(1{,}440 \text{ min/d}) = 421.85 \text{ lb/d}$$

Determine the number of pounds used in 30 days.

(421.85 lb/d)(30 d/month) = 12,655.5 lb/month, round to **12,700 lb/month of lime**

To find the dosage, use the "pounds" equation and rearrange to solve for dosage.

The equation is:

$$\textbf{Number, lb/d} = \textbf{(mgd)(Dosage, mg/L)(8.34 lb/gal)}$$

$$\text{Dosage, mg/L} = \frac{\text{lb/d}}{(\text{mgd})(8.34 \text{ lb/gal})}$$

$$\text{Lime dosage, mg/L} = \frac{421.85 \text{ lb/d}}{(4.212 \text{ mgd})(8.34 \text{ lb/d})}$$

= 12.01 mg/L, round to **12.0 mg/L of lime**

49. A plant is treating water at 17.25 mgd. If lime is being added at a rate of 410 g/min, what are the lime usage in pounds per day and the dosage in milligrams per liter?

First, find the pounds per day of lime usage.

$$\textbf{Pounds per day, lime} = \textbf{(g/min)(1,440 min/d)(1 lb/454 g)}$$

$$\text{Pounds per day, lime} = \text{(410 g/min)(1,440 min/d)(1 lb/454 g)} = \text{1,300 lb/d of lime}$$

Then, using the "pounds" equation, calculate the dosage in milligrams per liter by rearranging the formula and solving for dosage.

$$\text{Dosage, mg/L} = \frac{\text{lb/d}}{\text{(mgd)(8.34 lb/gal)}}$$

$$\text{Dosage, mg/L} = \frac{\text{1,300 lb/d}}{\text{(17.25 mgd)(8.34 lb/gal)}} = \textbf{9.0 mg/L of lime}$$

50. The dosage for alum at a treatment plant is 320 mL/min. The raw water flow rate is 1,240 gpm. If the flow rate is adjusted to 1,600 gpm, what should the theoretical alum dosage be in milliliters per minute, if everything else remains the same?

$$\frac{\text{320 mL/min}}{\text{1,240 gpm}} = \frac{x\text{, mL/min}}{\text{1,600 gpm}}$$

Solve for x.

$$x\text{, mL/min} = \frac{\text{(320 mL/min)(1,600 gpm)}}{\text{1,240 gpm}} = \text{412.9 mL/min, round to } \textbf{410 mL/min}$$

51. A treatment plant is adding 408.33 g/min of soda ash to its treated water. If the plant is producing water at 32.2 mgd, what are the soda ash usage in pounds per day and dosage in milligrams per liter?

First, find the pounds per day of soda ash usage.

Pounds per day, soda ash = (g/min)(1,440 min/d)(1 lb/454 g)

Pounds per day, soda ash = (408.33 g/min)(1,440 min/d)(1 lb/454 g)
= **1,295.1 lb/d of soda ash**

Then, using the "pounds" equation, calculate the dosage in milligrams per liter.

$$\text{Dosage, mg/L} = \frac{1{,}295.1 \text{ lb/d}}{(32.2 \text{ mgd})(8.34 \text{ lb/gal})} = \mathbf{4.82 \text{ mg/L of soda ash}}$$

52. A well is to be disinfected with 62% calcium hypochlorite. The well is 276 ft in depth and 1.0 ft in diameter. Depth to water from the top of the casing is 103 ft. If the desired dose is 50.0 mg/L, how many pounds of calcium hypochlorite are required?

First, find the length (in feet) of the water-filled casing.

Length of water-filled casing = Depth of well – Depth of water to top of casing

Length of water-filled casing = 276 ft – 103 ft = 173 ft

Then, determine the volume in gallons of water in the well casing using the following formula:

Volume, gal = (0.785)(Diameter)^2(Length)(7.48 gal/ft^3)

Volume, gal = (0.785)(1.0 ft)(1.0 ft)(173 ft)(7.48 gal/ft^3) = 1,015.8 gal

Next, determine the number of million gallons.

Million gallons = (1,015.8 gal)(1 M/1,000,000) = 0.0010158 mil gal

Last, using the "pounds" equation, calculate the number of pounds of calcium hypochlorite.

$$\text{Pounds, calcium hypochlorite} = \frac{(0.0010158 \text{ mil gal})(50.0 \text{ mg/L})(8.34 \text{ lb/gal})}{(62\% \text{ Available chlorine}/100\%)}$$

Pounds, calcium hypochlorite = 0.683 lb, round to **0.68 lb**

53. A fluoride dose of 1.10 mg/L is needed to treat a flow of 2,800 gpm. How many pounds per day of sodium fluorosilicate with a commercial purity of 98% and a fluoride ion content of 60.6% will be required? The water being treated contains 0.37 mg/L fluoride.

First, determine how many million gallons per day are being treated.

Million gallons per day = (2,800 gpm)(1,440 min/d)(1 M/1,000,000) = 4.032 mgd

Because there is natural fluoride (F) present, subtract the natural from the desired to get the dose required.

F dose required = 1.10 mg/L F – 0.37 mg/L natural F content = 0.73 mg/L F

Write the "pounds" equation with the addition of the percent purity and fluoride ion (F) content.

$$\textbf{Pounds per day, F compound} = \frac{\textbf{(mgd)(Dosage, mg/L)(8.34 lb/d)}}{\textbf{(\% purity/100\%)(\% F content/100\%)}}$$

$$\text{Pounds per day, F compound} = \frac{(4.032 \text{ mgd})(0.73 \text{ mg/L})(8.34 \text{ lb/d})}{(98\% \text{ purity}/100\%)(60.6\% \text{ F content}/100\%)}$$

Pounds per day, F compound = 41.33 lb/d, round to **41 lb/d of fluoride**

54. A polymer solution weighs 16.34 g per 10.0 mL. The water treatment plant is using 35 mL/min of the polymer for treating 24 ft^3/s. What is the polymer dosage?

First, determine the number of grams per milliliter.

$$16.34 \text{ g}/10.0 \text{ mL} = 1.634 \text{ g/mL}$$

Then, find the number of pounds per gallon for the polymer.

$$(1.634 \text{ g/mL})(3{,}785 \text{ mL/gal})(1 \text{ lb}/454 \text{ g}) = 13.62 \text{ lb/gal}$$

Next, convert cubic feet per second to million gallons per day.

$$(24 \text{ ft}^3\text{/s})(86{,}400 \text{ s/d})(7.48 \text{ gal/ft}^3)(1 \text{ M}/1{,}000{,}000) = 15.51 \text{ mgd}$$

Now, find the number of pounds per day of polymer usage.

$$(35 \text{ mL/min})(1{,}440 \text{ min/d})(1 \text{ gal}/3{,}785 \text{ mL})(13.62 \text{ lb/gal}) = 181.36 \text{ lb/d}$$

Last, calculate the dosage.

$$\textbf{Dosage, mg/L} = \frac{\textbf{lb/d}}{\textbf{(mgd)(8.34 lb/gal)}}$$

$$\text{Dosage, mg/L} = \frac{181.36 \text{ lb/d}}{(15.51 \text{ mgd})(8.34 \text{ lb/gal})} = \textbf{1.4 mg/L of polymer}$$

55. A treatment plant is adding 321 g/min of soda ash to its treated water. If the plant is producing water at 18.0 mgd, what is the soda ash usage in pounds per day and what is the dosage in milligrams per liter?

First, find the pounds per day of soda ash usage.

$$\textbf{Pounds per day, soda ash} = \textbf{(g/min)(1,440 min/d)(1 lb/454 g)}$$

$$\text{Pounds per day, soda ash} = (321 \text{ g/min})(1{,}440 \text{ min/d})(1 \text{ lb}/454 \text{ g})$$

$$= 1{,}018.15 \text{ lb/d, round to } \textbf{1,020 lb/d of soda ash}$$

Then, using the "pounds" equation, calculate the dosage in milligrams per liter.

$$\text{Dosage, mg/L} = \frac{1{,}018.15 \text{ lb/d}}{(18.0 \text{ mgd})(8.34 \text{ lb/gal})} = \textbf{6.78 mg/L of soda ash}$$

56. An operator was inspecting a tank (containing 34,827 L of water) from the access point on top of the tank, when a container of standard water solution consisting of 35.0 mg of iron dissolved in 50 mL fell out of the operator's pocket. The sample fell into the tank and mixed with the water. The iron was evenly dispersed throughout the tank. How many milligrams per liter of iron were added to the tank? Iron tests performed every month over the previous year indicated that processed water contained an average of 0.015 mg/L iron. Should the operator be concerned?

First, find the number of liters.

$$\frac{(34{,}827 \text{ L}) + 50 \text{ mL}}{(1 \text{ L}/1{,}000 \text{ mL})} = 34{,}827.05 \text{ L}$$

Next, determine the milligrams per liter of iron added to the tank.

$$\frac{35.0 \text{ mg}}{34{,}827 \text{ L}} = 0.0010 \text{ mg/L of iron}$$

Finally, calculate the total iron content to determine if the operator should be concerned.

$$\text{Iron, total mg/L} = 0.015 \text{ mg/L} + 0.0010 \text{ mg/L} = \mathbf{0.016\ mg/L\ of\ iron}$$

The operator should not be concerned about the iron, because the secondary maximum contaminant level for iron is 0.3 mg/L.

RATIO CALCULATIONS

57. A raw water flow of 25 ft³/s is prechlorinated with 450 lb of chlorine gas. If the flow is changed to 30 ft³/s, what should the adjustment to the chlorinator be?

$$\frac{\text{lb } Cl_2}{\text{flow}_1} = \frac{x \text{ lb } Cl_2}{\text{flow}_2}$$

Solve for x.

$$x \text{ lb } Cl_2 = \frac{(450 \text{ lb})(30 \text{ ft}^3\text{/s})}{25 \text{ ft}^3\text{/s}} = \mathbf{540\ lb\ of\ Cl_2}$$

58. The alum dosage for a plant with a flow of 15 ft³/s is 720 mL/min. If the raw water flow rate is adjusted to 12 ft³/s, what should the theoretical alum dosage in milliliters per minute be if everything else remains the same?

$$\frac{720 \text{ mL/min}}{15 \text{ ft}^3\text{/s}} = \frac{x, \text{ mL/min}}{12 \text{ ft}^3\text{/s}}$$

Solve for x.

$$x, \text{ mL/min} = \frac{(720 \text{ mL/min})(12 \text{ ft}^3\text{/s})}{15 \text{ ft}^3\text{/s}} = 576 \text{ mL/min, round to } \mathbf{580 \text{ mL/min}}$$

59. The dosage for polymer aid at a conventional filtration plant is 10.0 mL/min. The raw water flow rate is 1,825 gpm. If the flow rate is adjusted to 2,050 gpm, what should the theoretical polymer aid dosage be in milliliters per minute?

$$\frac{10.0 \text{ mL/min}}{1{,}825 \text{ gpm}} = \frac{x, \text{ mL/min}}{2{,}050 \text{ gpm}}$$

Solve for x.

$$x, \text{ mL/min} = \frac{(10.0 \text{ mL/min})(2{,}050 \text{ gpm})}{1{,}825 \text{ gpm}} = \mathbf{11.2 \text{ mL/min}}$$

FILTRATION RATE PROBLEMS

60. Four filters have a surface area of 760 ft each. What is the filtration rate in gallons per minute if they receive a total flow of 23.0 ft³/s?

First, convert the number of cubic feet per second to gallons per minute.

$$(23.0 \text{ ft}^3\text{/s})(7.48 \text{ gal/ft}^3)(60 \text{ s/min}) = 10{,}322.4 \text{ gpm}$$

Next, calculate the total surface area for all four filters.

$$(760 \text{ ft}^2)(4 \text{ filters}) = 3{,}040 \text{ ft}^2$$

$$\textbf{Filtration rate} = \frac{\textbf{Flow rate, gpm}}{\textbf{Filter surface area, ft}^2}$$

$$\text{Filtration rate} = \frac{10{,}322.4 \text{ gpm}}{3{,}040 \text{ ft}^2} = 3.396 \text{ gpm/ft}^2\text{, round to } \mathbf{3.4 \text{ gpm/ft}^2}$$

61. Four filters have a surface area of 840 ft each, measured to the nearest foot. What is the filtration rate in gallons per minute if they receive a flow of 44 ft^3/s?

First, convert the number of cubic feet per second to gallons per minute.

$$(44\ ft^3/s)(7.48\ gal/ft^3)(60\ s/min) = 19{,}747.2\ gpm$$

Next, determine the combined filtration area of the four filters.

$$\text{Total filter area} = (4\ \text{filters})(840\ ft) = 3{,}360\ ft$$

$$\textbf{Filtration rate} = \frac{\textbf{Flow rate, gpm}}{\textbf{Filter surface area, ft}^2}$$

$$\text{Filtration rate} = \frac{19{,}747.2\ gpm}{3{,}360\ ft^2} = \mathbf{5.9\ gpm/ft^2}$$

62. A water treatment plant has four filters with an average flow rate of 4.8 gpm/ft^2. If the plant flow is 21.4 ft^3/s, what is the filtration area of each filter?

First, calculate the number of gallons per minute.

$$(21.4\ ft^3/s)(7.48\ gal/ft^3)(60\ s/min) = 9{,}604.32\ gpm$$

$$\textbf{Filtration rate} = \frac{\textbf{Flow rate, gpm}}{\textbf{Filter surface area, ft}^2}$$

Rearrange the formula.

$$\text{Filter surface area, ft}^2 = \frac{(\text{Flow rate, gpm})}{(\text{Filtration rate})}$$

$$\text{Filter surface area, ft}^2 = \frac{(9{,}604.32\ gpm)}{(4.8\ gpm/ft^2)} = 2{,}000.9\ ft^2\ \text{(for all four filters)}$$

Filter area for each filter = 2,000.9 ft^2/4 filters = 500.225 ft^2/filter, round to **500 ft^2/filter**

63. A treatment plant produced a total of 40.3 mil gal in 3 days. All six of the filters were backwashed once during this time. Each filter is 32 ft by 24 ft. What is the unit filter run volume (UFRV) in gallons per square foot for the last filter backwashed, assuming uniform distribution of water across all filters during the three days?

First, find the number of gallons produced by each filter.

$$\text{Filter production, gal} = \frac{(40.3 \text{ mil gal})(1{,}000{,}000/1 \text{ M})}{6 \text{ filters}} = 6{,}716{,}667 \text{ gal}$$

$$\mathbf{UFRV, gal/ft^2} = \frac{\mathbf{Total\ gallons\ filtered}}{\mathbf{Filter\ surface\ area,\ ft^2}}$$

$$\text{UFRV, gal/ft}^2 = \frac{6{,}716{,}667 \text{ gal}}{(32 \text{ ft})(24 \text{ ft})} = 8{,}746 \text{ gal/ft}^2\text{, round to } \mathbf{8{,}700\ gal/ft^2}$$

64. A water treatment plant has eight filters with an average flow rate of 6.43 gpm/ft². If the plant flow is 86 ft³/s, what is the area of each filter?

First, calculate the number of gallons per minute.

$$(86 \text{ ft}^3\text{/s})(7.48 \text{ gal/ft}^3)(60 \text{ s/min}) = 38{,}596.8 \text{ gpm}$$

$$\mathbf{Filtration\ rate} = \frac{\mathbf{Flow\ rate,\ gpm}}{\mathbf{Filter\ surface\ area,\ ft^2}}$$

Rearrange the formula.

$$\text{Filter surface area, ft}^2 = \frac{(\text{Flow rate, gpm})}{(\text{Filtration rate})}$$

$$\text{Filter surface area, ft}^2 = \frac{(38{,}596.8 \text{ gpm})}{(6.43 \text{ gpm/ft}^2)} = 6{,}002.6 \text{ ft}^2 \text{ (for all eight filters)}$$

$$\text{Filter area for each filter} = 6{,}002.6 \text{ ft}^2\text{/8 filters} = \mathbf{750\ ft^2\ for\ each\ filter}$$

65. A treatment plant produced a total of 56 mil gal in 4 days during which all eight filters were backwashed once each. Each filter is 35 ft by 25 ft. What is the UFRV in gallons per square foot for the last filter to be backwashed (assume uniform filtration rate across all eight filters for the 4 days)?

First, find the number of gallons produced by each filter.

$$\text{Filter production, gal} = \frac{(56\text{ mil gal})(1{,}000{,}000/1\text{ M})}{8\text{ filters}} = 7{,}000{,}000\text{ gal}$$

$$\textbf{UFRV, gal/ft}^2 = \frac{\textbf{Total gallons filtered}}{\textbf{Filter surface area, ft}^2}$$

$$\text{UFRV, gal/ft}^2 = \frac{7{,}000{,}000\text{ gal}}{(35\text{ ft})(25\text{ ft})} = \mathbf{8{,}000\text{ gal/ft}^2}$$

66. A filter has a surface area of 750 ft. What is the filtration rate in gallons per minute if it receives a flow of 6.0 ft³/s?

First, convert the number of cubic feet per second to gallons per minute.

$$(6.0\text{ ft}^3/\text{s})(7.48\text{ gal/ft}^3)(60\text{ s/min}) = 2{,}692.8\text{ gpm}$$

$$\textbf{Filtration rate} = \frac{\textbf{Flow rate, gpm}}{\textbf{Filter surface area, ft}^2}$$

$$\text{Filtration rate} = \frac{2{,}692.8\text{ gpm}}{750\text{ ft}^2} = 3.59\text{ gpm/ft}^2\text{, round to } \mathbf{3.6\text{ gpm/ft}^2}$$

67. A treatment plant's filter produced a total of 3.0 mil gal between backwashes. The filter is 32 ft by 22 ft. What is the UFRV in gallons per square foot?

First, find the number of gallons produced by each filter.

$$\text{Filter production, gal} = (3.0 \text{ mil gal})(1{,}000{,}000/1 \text{ M}) = 3{,}000{,}000 \text{ gal}$$

$$\textbf{UFRV, gal/ft}^2 = \frac{\textbf{Total gallons filtered}}{\textbf{Filter surface area, ft}^2}$$

$$\text{UFRV, gal/ft}^2 = \frac{3{,}000{,}000 \text{ gal}}{(32 \text{ ft})(22 \text{ ft})} = 4{,}261 \text{ gal/ft}^2\text{, round to } \mathbf{4{,}300 \text{ gal/ft}^2}$$

68. A treatment plant produced a total of 36.4 mil gal in 8 days. Eight filters were backwashed once each during this time. To the nearest foot, each filter is 30 ft by 24 ft. What is the UFRV in gallons per square foot for the last filter to be backwashed (assume uniform filtration rate across all eight filters for the 4 days)? *Note*: Give answer to two significant figures.

First, find the number of gallons produced by each filter.

$$\text{Filter production, gal} = \frac{(36.4 \text{ mil gal})(1{,}000{,}000/1 \text{ M})}{8 \text{ filters}} = 4{,}550{,}000 \text{ gal}$$

$$\textbf{UFRV, gal/ft}^2 = \frac{\textbf{Total gallons filtered}}{\textbf{Filter surface area, ft}^2}$$

$$\text{UFRV, gal/ft}^2 = \frac{4{,}550{,}000 \text{ gal}}{(30 \text{ ft})(24 \text{ ft})} = 6{,}319.44 \text{ gal/ft}^2\text{, round to } \mathbf{6{,}300 \text{ gal/ft}^2}$$

69. A water treatment plant has six filters with an average flow rate of 5.89 gpm/ft². If the plant flow is 63 ft³/s, what is the area of each filter?

First, calculate the number of gallons per minute.

$$(63 \text{ ft}^3/\text{s})(7.48 \text{ gal/ft}^3)(60 \text{ s/min}) = 28{,}274 \text{ gpm}$$

$$\textbf{Filtration rate} = \frac{\textbf{Flow rate, gpm}}{\textbf{Filter surface area, ft}^2}$$

Rearrange the formula.

$$\text{Filter surface area, ft}^2 = \frac{(\text{Flow rate, gpm})}{(\text{Filtration rate})}$$

$$\text{Filter surface area, ft}^2 = \frac{(28{,}274\ \text{gpm})}{(5.89\ \text{gpm/ft}^2)} = 4{,}800.4\ \text{ft}^2\text{, for all six filters}$$

$$\text{Filter area for each filter} = 4{,}800.4\ \text{ft}^2/6\ \text{filters} = \mathbf{800\ ft^2\ for\ each\ filter}$$

BACKWASH RATE PROBLEMS

70. A filter has an area of 780 ft² with a backwash (BW) pumping rate of 15 ft³/s. What is the BW rate in gallons per minute per square foot?

The equation is:

$$\textbf{BW rate, gpm/ft}^2 = \frac{\textbf{BW pumping rate, gpm}}{\textbf{Filter area, ft}^2}$$

Then, convert the pumping rate in cubic feet per second to gallons per minute.

$$\text{Gallons per minute} = (15\ \text{ft}^3\text{/s})(7.48\ \text{gal/ft}^3)(60\ \text{s/min}) = 6{,}732\ \text{gpm}$$

$$\text{BW rate, gpm/ft}^2 = \frac{6{,}732\ \text{gpm}}{780\ \text{ft}^2} = 8.63\ \text{gpm/ft}^2\text{, round to } \mathbf{8.6\ gpm/ft^2}$$

71. What is the backwash (BW) rate in gallons per minute per square foot if a filter has an area of 780 ft² with a BW pumping rate of 12.5 ft³/s?

The equation is:

$$\textbf{BW rate, gpm/ft}^2 = \frac{\textbf{BW pumping rate, gpm}}{\textbf{Filter area, ft}^2}$$

Then, convert the pumping rate in cubic feet per second to gallons per minute.

$$\text{Gallons per minute} = (12.5\ \text{ft}^3\text{/s})(7.48\ \text{gal/ft}^3)(60\ \text{s/min}) = 5{,}610\ \text{gpm}$$

$$\text{BW rate, gpm/ft}^2 = \frac{5{,}610\ \text{gpm}}{780\ \text{ft}^2} = 7.19\ \text{gpm/ft}^2\text{, round to } \mathbf{7.2\ gpm/ft^2}$$

72. What is the backwash (BW) rate in gallons per minute per square foot if a filter has an area of 750 ft² with a BW pumping rate of 13.25 ft³/s?

The equation is:

$$\textbf{BW rate, gpm/ft}^2 = \frac{\textbf{BW pumping rate, gpm}}{\textbf{Filter area, ft}^2}$$

Then, convert the pumping rate in cubic feet per second to gallons per minute.

$$\text{Gallons per minute} = (13.25\ \text{ft}^3/\text{s})(7.48\ \text{gal/ft}^3)(60\ \text{s/min}) = 5{,}946.6\ \text{gpm}$$

$$\text{BW rate, gpm/ft}^2 = \frac{5{,}946.6\ \text{gpm}}{750\ \text{ft}^2} = 7.93\ \text{gpm/ft}^2\text{, round to } \mathbf{7.9\ gpm/ft^2}$$

73. What is the backwash (BW) rate in gallons per minute per square foot if a filter has an area of 580 ft² with a BW pumping rate of 11.74 ft³/s?

The equation is:

$$\textbf{BW rate, gpm/ft}^2 = \frac{\textbf{BW pumping rate, gpm}}{\textbf{Filter area, ft}^2}$$

Then, convert the pumping rate in cubic feet per second to gallons per minute.

$$\text{Gallons per minute} = (11.74\ \text{ft}^3/\text{s})(7.48\ \text{gal/ft}^3)(60\ \text{s/min}) = 5{,}268.9\ \text{gpm}$$

$$\text{BW rate, gpm/ft}^2 = \frac{5{,}268.9\ \text{gpm}}{580\ \text{ft}^2} = 9.08\ \text{gpm/ft}^2\text{, round to } \mathbf{9.1\ gpm/ft^2}$$

74. What is the backwash (BW) rate in gallons per minute per square foot if a filter has an area of 600 ft² with a BW pumping rate of 13 ft³/s?

The equation is:

$$\textbf{BW rate, gpm/ft}^2 = \frac{\textbf{BW pumping rate, gpm}}{\textbf{Filter area, ft}^2}$$

Then, convert the pumping rate in cubic feet per second to gallons per minute.

$$\text{Gallons per minute} = (13\ \text{ft}^3/\text{s})(7.48\ \text{gal/ft}^3)(60\ \text{s/min}) = 5{,}834.4\ \text{gpm}$$

$$\text{BW rate, gpm/ft}^2 = \frac{5{,}834.4\ \text{gpm}}{600\ \text{ft}^2} = 9.724\ \text{gpm/ft}^2\text{, round to } \mathbf{10\ gpm/ft^2}$$

75. What is the backwash (BW) pumping rate if the desired BW rate is 6.0 gpm/ft² for a filter that is 40 ft by 32 ft, measured to the nearest foot? Assume two significant figures.

The formula for the BW pumping rate can be obtained by rearranging the formula in the last problem.

$$\textbf{BW pumping rate, gpm} = \textbf{(BW rate, gpm/ft}^2\textbf{)(Filter area, ft}^2\textbf{)}$$

Then, find the surface area of the filter.

$$\text{Area} = \text{L} \times \text{W} = (40 \text{ ft})(32 \text{ ft}) = 1{,}280 \text{ ft}^2$$

$$\text{BW pumping rate} = (6.0 \text{ gpm/ft}^2)(1{,}280 \text{ ft}^2) = 7{,}680 \text{ gpm, round to } \mathbf{7{,}700 \text{ gpm}}$$

76. What is the backwash (BW) rate in gallons per minute per square foot if a filter has an area of 620 ft² with a BW pumping rate of 14 ft³/s?

The equation is:

$$\textbf{BW rate, gpm/ft}^2 = \frac{\textbf{BW pumping rate, gpm}}{\textbf{Filter area, ft}^2}$$

Then, convert the pumping rate in cubic feet per second to gallons per minute.

$$\text{Gallons per minute} = (14 \text{ ft}^3\text{/s})(7.48 \text{ gal/ft}^3)(60 \text{ s/min}) = 6{,}283.2 \text{ gpm}$$

$$\text{BW rate, gpm/ft}^2 = \frac{6{,}283.2 \text{ gpm}}{620 \text{ ft}^2} = 10.13 \text{ gpm/ft}^2\text{, round to } \mathbf{10 \text{ gpm/ft}^2}$$

CORROSION PROBLEMS

77. Test results of distribution water give a pH of 7.2 and a pH_s of 7.4. What is the Langelier index (LI) and what does this tell you about the distribution system in the area where the water was collected?

The equation is:

$$\textbf{LI} = \textbf{pH} - \textbf{pH}_s$$

$$\text{LI} = 7.2 - 7.4 = \mathbf{-0.2}$$

A negative LI indicates the water is corrosive.

78. Test results of distribution water give a pH of 7.8 and a pH_s of 7.6. What is the Langelier index (LI) and what does this tell you about the distribution system in the area where the water was collected?

The equation is:

$$\mathbf{LI = pH - pH_s}$$

$$LI = 7.8 - 7.6 = \mathbf{0.2}$$

A positive LI indicates the water is scale forming.

79. Test results of distribution water give a pH of 7.94 and a pH_s of 7.62. What is the Langelier index (LI) and what does this tell you about the distribution system in the area where the water was collected?

The equation is:

$$\mathbf{LI = pH - pH_s}$$

$$LI = 7.94 - 7.62 = \mathbf{0.32}$$

A positive LI indicates the water is scale forming.

80. Test results of distribution water give a pH of 7.8 and a pH_s of 7.5. What is the Langelier index (LI) and what does this tell you about the distribution system in the area where the water was collected?

The equation is:

$$\mathbf{LI = pH - pH_s}$$

$$LI = 7.8 - 7.5 = \mathbf{0.3}$$

A positive LI indicates the water is scale forming.

SOFTENING CALCULATIONS

81. What is the hardness in milligrams per liter of a well's water if the hardness is 22.5 grains per gallon (gpg)?

The equation is:

$$\textbf{Hardness, mg/L} = \frac{\textbf{(Hardness, gpg)(17.12 mg/L)}}{\textbf{1 gpg}}$$

$$\text{Hardness, mg/L} = \frac{(22.5\text{ gpg})(17.12\text{ mg/L})}{1\text{ gpg}} = \textbf{385 mg/L}$$

82. If the hardness of a water sample is 14 gpg (grains per gallon), what is its hardness in milligrams per liter?

The equation is:

$$\textbf{Hardness, mg/L} = \frac{\textbf{(Hardness, gpg)(17.12 mg/L)}}{\textbf{1 gpg}}$$

$$\text{Hardness, mg/L} = \frac{(14\text{ gpg})(17.12\text{ mg/L})}{1\text{ gpg}} = 239.68\text{ mg/L, round to } \textbf{240 mg/L}$$

83. What is the hardness in milligrams per liter of a treatment plant's water if the hardness is 17.45 gpg (grains per gallon)?

The equation is:

$$\textbf{Hardness, mg/L} = \frac{\textbf{(Hardness, gpg)(17.12 mg/L)}}{\textbf{1 gpg}}$$

$$\text{Hardness, mg/L} = \frac{(17.45\text{ gpg})(17.12\text{ mg/L})}{1\text{ gpg}} = \textbf{298.7 mg/L}$$

84. Water from a well has a total hardness of 19.8 gpg (grains per gallon). What is the concentration, if expressed as milligrams per liter?

The equation is:

$$\textbf{Hardness, mg/L} = \frac{\textbf{(Hardness, gpg)(17.12 mg/L)}}{\textbf{1 gpg}}$$

$$\text{Hardness, mg/L} = \frac{(19.8 \text{ gpg})(17.12 \text{ mg/L})}{1 \text{ gpg}} = \textbf{339 mg/L}$$

85. How many grains per gallon does a well's water have if its hardness is 300 mg/L?

The equation is:

$$\textbf{Hardness, mg/L} = \frac{\textbf{(Hardness, gpg)(17.12 mg/L)}}{\textbf{1 gpg}}$$

Rearrange the equation.

$$\text{Hardness, gpg} = \frac{(\text{Hardness, mg/L})(1 \text{ gpg})}{17.12 \text{ mg/L}}$$

$$\text{Hardness, gpg} = \frac{(300 \text{ mg/L})(1 \text{ gpg})}{17.12 \text{ mg/L}} = 17.52 \text{ gpg, round to } \textbf{20 gpg}$$

86. Water from a well has a total hardness of 24.1 gpg (grains per gallon). What is the concentration if expressed as milligrams per liter?

The equation is:

$$\textbf{Hardness, mg/L} = \frac{\textbf{(Hardness, gpg)(17.12 mg/L)}}{\textbf{1 gpg}}$$

$$\text{Hardness, mg/L} = \frac{(24.1 \text{ gpg})(17.12 \text{ mg/L})}{1 \text{ gpg}} = \textbf{413 mg/L}$$

87. Water from a well sample has 72 mg/L of magnesium (Mg) as $CaCO_3$ and 112 mg/L of calcium (Ca) as $CaCO_3$. What is the total hardness of the sample as $CaCO_3$?

The equation is:

$$\textbf{Total hardness} = \textbf{Ca hardness, mg/L as } \mathbf{CaCO_3} + \textbf{Mg hardness, mg/L as } \mathbf{CaCO_3}$$

$$\text{Total hardness} = 112 \text{ mg/L Ca, as } CaCO_3 + 72 \text{ mg/L Mg, as } CaCO_3$$

$$= \mathbf{184\ mg/L,\ as\ CaCO_3}$$

88. Water from a well has a total hardness of 20.2 gpg (grains per gallon). What is the concentration if expressed as milligrams per liter?

The equation is:

$$\textbf{Hardness, mg/L} = \frac{\textbf{(Hardness, gpg)(17.12 mg/L)}}{\textbf{1 gpg}}$$

$$\text{Hardness, mg/L} = \frac{(20.2 \text{ gpg})(17.12 \text{ mg/L})}{1 \text{ gpg}} = 345.8 \text{ mg/L, round to } \mathbf{346\ mg/L}$$

89. How many grains per gallon does a particular water have if its hardness is 285 mg/L?

The equation is:

$$\textbf{Hardness, mg/L} = \frac{\textbf{(Hardness, gpg)(17.12 mg/L)}}{\textbf{1 gpg}}$$

Rearrange the equation.

$$\text{Hardness, gpg} = \frac{(\text{Hardness, mg/L})(1 \text{ gpg})}{17.12 \text{ mg/L}}$$

$$\text{Hardness, gpg} = \frac{(285 \text{ mg/L})(1 \text{ gpg})}{17.12 \text{ mg/L}} = \mathbf{16.6\ gpg}$$

CALCIUM HARDNESS CALCULATIONS

Equivalent Weights

Use the following equivalent weights for the problems that follow:

Constituent		Equivalent Weight
Calcium (Ca)	=	20.04
Calcium carbonate, $CaCO_3$	=	50.045
Magnesium (Mg)	=	12.15

90. What is the calcium hardness, as $CaCO_3$, if the water sample has a calcium content of 67 mg/L?

The equation is:

$$\frac{\textbf{Ca hardness, mg/L as CaCO}_3}{\textbf{Equivalent weight of CaCO}_3} = \frac{\textbf{Ca, mg/L}}{\textbf{Equivalent weight of Ca}}$$

Substitute known values.

$$\frac{\text{Ca hardness, mg/L as CaCO}_3}{50.045} = \frac{67 \text{ mg/L Ca}}{20.04}$$

Rearrange the equation.

$$\text{Calcium hardness, mg/L as CaCO}_3 = \frac{(67 \text{ mg/L Ca})(50.045)}{20.04}$$

$$\text{Calcium hardness, mg/L as CaCO}_3$$

$$= 167.32 \text{ mg/L, round to } \textbf{170 mg/L calcium as CaCO}_3$$

91. What is the magnesium hardness, as $CaCO_3$, if the water sample has a magnesium content of 29 mg/L?

The equation is:

$$\frac{\textbf{Mg hardness, mg/L as CaCO}_3}{\textbf{Equivalent weight of CaCO}_3} = \frac{\textbf{Mg, mg/L}}{\textbf{Equivalent weight of Mg}}$$

Substitute known values.

$$\frac{\textbf{Mg hardness, mg/L as CaCO}_3}{\textbf{50.045}} = \frac{\textbf{29 mg/L Ca}}{\textbf{12.15}}$$

Rearrange and solve.

$$\text{Magnesium hardness, mg/L as CaCO}_3 = \frac{(29\text{ mg/L Ca})(50.045)}{12.15}$$

$$\text{Magnesium hardness, mg/L as CaCO}_3$$

$$= 119.45\text{ mg/L, round to } \textbf{120 mg/L magnesium, as CaCO}_3$$

92. What is the calcium hardness, as $CaCO_3$, if the water sample has a calcium content of 99 mg/L?

The equation is:

$$\frac{\textbf{Ca hardness, mg/L as CaCO}_3}{\textbf{Equivalent weight of CaCO}_3} = \frac{\textbf{Ca, mg/L}}{\textbf{Equivalent weight of Ca}}$$

Substitute known values.

$$\frac{\text{Ca hardness, mg/L as CaCO}_3}{50.045} = \frac{99\text{ mg/L Ca}}{20.04}$$

Rearrange the equation.

$$\text{Calcium hardness, mg/L as CaCO}_3 = \frac{(99\text{ mg/L Ca})(50.045)}{20.04}$$

$$\text{Calcium hardness, mg/L as CaCO}_3$$

$$= 247.23\text{ mg/L, round to } \textbf{250 mg/L calcium as CaCO}_3$$

93. What is the calcium hardness, as $CaCO_3$, if the water sample has a calcium content of 84 mg/L?

The equation is:

$$\frac{\textbf{Ca hardness, mg/L as CaCO}_3}{\textbf{Equivalent weight of CaCO}_3} = \frac{\textbf{Ca, mg/L}}{\textbf{Equivalent weight of Ca}}$$

Substitute known values.

$$\frac{\text{Ca hardness, mg/L as CaCO}_3}{50.045} = \frac{84 \text{ mg/L Ca}}{20.04}$$

Rearrange the equation.

$$\text{Calcium hardness, mg/L as CaCO}_3 = \frac{(84 \text{ mg/L Ca})(50.045)}{20.04}$$

$$\text{Calcium hardness, mg/L as CaCO}_3$$

$$= 209.77 \text{ mg/L, round to } \textbf{210 mg/L calcium as CaCO}_3$$

94. What is the calcium hardness, as $CaCO_3$, if the water sample has a calcium content of 105 mg/L?

The equation is:

$$\frac{\textbf{Ca hardness, mg/L as CaCO}_3}{\textbf{Equivalent weight of CaCO}_3} = \frac{\textbf{Ca, mg/L}}{\textbf{Equivalent weight of Ca}}$$

Substitute known values and rearrange to solve.

$$\text{Calcium hardness, mg/L as CaCO}_3 = \frac{(105 \text{ mg/L Ca})(50.045)}{20.04}$$

$$\text{Calcium hardness, mg/L as CaCO}_3$$

$$= 262.21 \text{ mg/L, round to } \textbf{262 mg/L calcium as CaCO}_3$$

95. What is the calcium hardness, as $CaCO_3$, if the water sample has a calcium content of 157 mg/L?

The equation is:

$$\frac{\textbf{Ca hardness, mg/L as CaCO}_3}{\textbf{Equivalent weight of CaCO}_3} = \frac{\textbf{Ca, mg/L}}{\textbf{Equivalent weight of Ca}}$$

Substitute known values and rearrange to solve.

$$\text{Calcium hardness, mg/L as CaCO}_3 = \frac{(157 \text{ mg/L Ca})(50.045)}{20.04}$$

$$= \textbf{392 mg/L calcium as CaCO}_3$$

96. What is the magnesium hardness, as $CaCO_3$, if the water sample has a magnesium content of 37 mg/L?

The equation is:

$$\frac{\textbf{Mg hardness, mg/L as CaCO}_3}{\textbf{Equivalent weight of CaCO}_3} = \frac{\textbf{Mg, mg/L}}{\textbf{Equivalent weight of Mg}}$$

Rearrange and solve.

$$\text{Magnesium hardness, mg/L as CaCO}_3 = \frac{(37 \text{ mg/L Ca})(50.045)}{12.15}$$

$$\text{Magnesium hardness, mg/L as CaCO}_3$$
$$= 152.4 \text{ mg/L, round to } \textbf{150 mg/L magnesium as CaCO}_3$$

97. What is the magnesium hardness, as $CaCO_3$, if the water sample has a magnesium content of 14 mg/L?

The equation is:

$$\frac{\textbf{Mg hardness, mg/L as CaCO}_3}{\textbf{Equivalent weight of CaCO}_3} = \frac{\textbf{Mg, mg/L}}{\textbf{Equivalent weight of Mg}}$$

$$\text{Magnesium hardness, mg/L as CaCO}_3 = \frac{(14 \text{ mg/L Ca})(50.045)}{12.15}$$

Magnesium hardness, mg/L as $CaCO_3$

= 57.67 mg/L, round to **57 mg/L magnesium as $CaCO_3$**

TOTAL HARDNESS AS EXPRESSED IN CARBONATE AND NONCARBONATE HARDNESS

98. Find the noncarbonate hardness of a water sample if the alkalinity is 123 mg/L as $CaCO_3$ and the total hardness is 158 mg/L.

The equation is:

Total hardness, mg/L as $CaCO_3$ = Carbonate hardness, mg/L as $CaCO_3$ + Noncarbonate hardness, mg/L as $CaCO_3$

Rearrange the equation to solve for noncarbonate hardness and substitute known quantities.

Noncarbonate hardness, mg/L as $CaCO_3$ = 158 mg/L – 123 mg/L
= **35 mg/L noncarbonate hardness as $CaCO_3$**

99. Find the noncarbonate hardness of a water sample if the alkalinity is 256 mg/L as $CaCO_3$ and the total hardness is 345 mg/L.

The equation is:

Total hardness, mg/L as $CaCO_3$ = Carbonate hardness, mg/L as $CaCO_3$ + Noncarbonate hardness, mg/L as $CaCO_3$

Because the total hardness is greater than the alkalinity, there will be both carbonate and noncarbonate hardness present in the sample. Rearrange the equation to solve for noncarbonate hardness and substitute known quantities.

Noncarbonate hardness, mg/L as $CaCO_3$ = 345 mg/L – 256 mg/L
= **89 mg/L noncarbonate hardness as $CaCO_3$**

100. What is the carbonate and noncarbonate hardness of a water sample containing 178 mg/L total hardness as $CaCO_3$ and an alkalinity content of 189 mg/L as $CaCO_3$?

Because the total hardness is less than the alkalinity, all the hardness will be carbonate hardness. It follows that noncarbonate hardness of this water is zero, and:

Total hardness, mg/L as $CaCO_3$ = Carbonate hardness, mg/L as $CaCO_3$

Total hardness, 178 mg/L as $CaCO_3$ = Carbonate hardness, mg/L as $CaCO_3$
= **178 mg/L of carbonate hardness**

noncarbonate hardness = 0

101. Find the total hardness of a water sample if the alkalinity is 109 mg/L as $CaCO_3$ and the noncarbonate hardness is 47 mg/L.

The equation is:

$$\textbf{Total hardness, mg/L as } CaCO_3 = \textbf{Carbonate hardness, mg/L as } CaCO_3 + \textbf{Noncarbonate hardness, mg/L as } CaCO_3$$

Rearrange the equation to solve for noncarbonate hardness and substitute known quantities.

$$\text{Total hardness, mg/L as } CaCO_3 = 47 \text{ mg/L} + 109 \text{ mg/L}$$
$$= \textbf{156 mg/L total hardness as } CaCO_3$$

102. Find the noncarbonate hardness of a water sample if the alkalinity is 220 mg/L as $CaCO_3$ and the total hardness is 308 mg/L.

The equation is:

$$\textbf{Total hardness, mg/L as } CaCO_3 = \textbf{Carbonate hardness, mg/L as } CaCO_3 + \textbf{Noncarbonate hardness, mg/L as } CaCO_3$$

Because the total hardness is greater than the alkalinity, there will be both carbonate and noncarbonate hardness present in the sample. As above, rearrange the equation to solve for noncarbonate hardness and substitute known quantities.

$$\text{Noncarbonate hardness, mg/L as } CaCO_3 = 308 \text{ mg/L} - 220 \text{ mg/L}$$
$$= \textbf{88 mg/L noncarbonate hardness as } CaCO_3$$

ION EXCHANGE SOFTENING PROBLEMS

103. A water softener has 52.0 ft^3 of cationic resin. The capacity of the resin is 35,000 grains/ft^3. How many gallons of water will be produced before the softener is exhausted if the water hardness is 213 mg/L?

First, find the hardness of the water in grains per gallon.

$$\frac{(213 \text{ mg/L})(1 \text{ gpg})}{17.12 \text{ mg/L}} = 12.44 \text{ gpg}$$

Then, find the exchange capacity.

$$\text{Exchange capacity} = (35{,}000 \text{ grains/ft}^3)(52.0 \text{ ft}^3) = 1{,}820{,}000 \text{ grains}$$

The equation is:

$$\textbf{Water treatment capacity, gal} = \frac{\textbf{Exchange capacity, grains}}{\textbf{Hardness, gpg}}$$

$$\text{Water treatment capacity, gal} = \frac{1{,}820{,}000 \text{ grains}}{12.44 \text{ gpg}} = 146{,}302 \text{ gal, round to } \textbf{150,000 gal}$$

104. The exchange capacity of a softener is 7,015,000 grains. The softener treats water with an average hardness of 318 mg/L. What is the capacity of the softener in gallons?

First, determine the hardness of the water in grains per gallon.

$$\frac{(318 \text{ mg/L})(1 \text{ gpg})}{17.12 \text{ mg/L}} = 18.57 \text{ gpg}$$

The equation is:

$$\textbf{Water treatment capacity, gal} = \frac{\textbf{Exchange capacity, grains}}{\textbf{Hardness, gpg}}$$

$$\text{Water treatment capacity, gal} = \frac{7{,}015{,}000 \text{ grains}}{18.57 \text{ gpg}} = 377{,}760 \text{ gal, round to } \textbf{378,000 gal}$$

105. An ion exchange softener is treating a flow rate of 280 gpm. What is the operating time, in hours, if the softener unit treats 604,000 gal before it requires regeneration?

The equation is:

$$\textbf{Operating time, hr} = \frac{\textbf{Treated water, gal}}{\textbf{Flow rate, gph}}$$

Convert flow rate from gallons per minute to gallons per hour.

$$(280 \text{ gpm})(60 \text{ min/hr}) = 16{,}800 \text{ gph}$$

$$\text{Operating time, hr} = \frac{604{,}000 \text{ gal}}{16{,}800 \text{ gph}} = 35.95 \text{ hr, round to } \mathbf{36\ hr}$$

106. Three softener units have 145 ft³ of resin with a capacity of 27.0 kilograins/ft³. How many gallons of water will the units treat if the water contains 15.9 gpg?

First, convert kilograins to grains.

$$(27.0 \text{ kilograins/ft}^3)(1{,}000 \text{ grains/kilograins}) = 27{,}000 \text{ grains/ft}^3$$

Next, find the capacity of the unit in grains.

$$(145 \text{ ft}^3)(27{,}000 \text{ grains/ft}^3) = 3{,}915{,}000 \text{ grains}$$

The equation is:

$$\textbf{Water treatment capacity, gal} = \frac{\textbf{Exchange capacity, grains}}{\textbf{Hardness, gpg}}$$

$$\text{Water treatment capacity, gal} = \frac{3{,}915{,}000 \text{ grains}}{15.9 \text{ gpg}}$$

$$= 246{,}226 \text{ gal, round to } \mathbf{246{,}000\ gal}$$

107. The exchange capacity of a softener is 4,525,000 grains. The softener treats water with an average hardness of 340 mg/L. What is the capacity of the softener in gallons?

First, determine the hardness of the water in grains per gallon by dividing by 17.12 mg/L per grains per gallon.

$$\frac{(340 \text{ mg/L})(1 \text{ gpg})}{(17.12 \text{ mg/L})} = 19.86 \text{ gpg}$$

The equation is:

$$\textbf{Water treatment capacity, gal} = \frac{\textbf{Exchange capacity, grains}}{\textbf{Hardness, gpg}}$$

$$\text{Water treatment capacity, gal} = \frac{4{,}525{,}000 \text{ grains}}{19.86 \text{ gpg}}$$

$$= 227{,}845 \text{ gal, round to } \textbf{230,000 gal}$$

108. A softener unit has 125 ft³ of resin with a capacity of 24 kilograins/ft³. How many gallons of water will the unit treat if the water contains 12.5 gpg?

First, convert kilograins to grains.

$$(24 \text{ kilograins/ft}^3)(1{,}000 \text{ grains/kilograins}) = 24{,}000 \text{ grains/ft}^3$$

Next, find the capacity of the unit in grains.

$$(125 \text{ ft}^3)(24{,}000 \text{ grains/ft}^3) = 3{,}000{,}000 \text{ grains}$$

The equation is:

$$\textbf{Water treatment capacity, gal} = \frac{\textbf{Exchange capacity, grains}}{\textbf{Hardness, gpg}}$$

$$\text{Water treatment capacity, gal} = \frac{3{,}000{,}000 \text{ grains}}{12.5 \text{ gpg}} = \textbf{240,000 gal}$$

109. A softener unit has 300.0 ft^3 of resin with a capacity of 25.4 kilograins/ft^3. How many gallons of water will the unit treat if the water contains 16.2 gpg?

First, convert kilograins to grains.

$$(25.4 \text{ kilograins/ft}^3)(1{,}000 \text{ grains/kilograins}) = 25{,}400 \text{ grains/ft}^3$$

Next, find the capacity of the unit in grains.

$$(300.0 \text{ ft}^3)(25{,}400 \text{ grains/ft}^3) = 7{,}620{,}000 \text{ grains}$$

The equation is:

$$\textbf{Water treatment capacity, gal} = \frac{\textbf{Exchange capacity, grains}}{\textbf{Hardness, gpg}}$$

$$\text{Water treatment capacity, gal} = \frac{7{,}620{,}000 \text{ grains}}{16.2 \text{ gpg}}$$

$$= 470{,}370 \text{ gal, round to } \mathbf{470{,}000 \text{ gal}}$$

110. An ion exchange softener is treating a flow rate of 125 gpm. What is the operating time, in hours, if the softener unit treats 425,500 gal before it requires regeneration?

The equation is:

$$\textbf{Operating time, hr} = \frac{\textbf{Treated water, gal}}{\textbf{Flow rate, gph}}$$

First, convert flow rate from gallons per minute to gallons per hour.

$$(125 \text{ gpm})(60 \text{ min/hr}) = 7{,}500 \text{ gph}$$

$$\text{Operating time, hr} = \frac{425{,}500 \text{ gal}}{7{,}500 \text{ gph}} = \mathbf{56.7 \text{ hr}}$$

111. An ion exchange softener is treating a flow rate of 200 gpm. What is the operating time, in hours, if the softener unit treats 386,400 gal before it requires regeneration?

The equation is:

$$\textbf{Operating time, hr} = \frac{\textbf{Treated water, gal}}{\textbf{Flow rate, gph}}$$

Convert flow rate from gallons per minute to gallons per hour.

$$(200 \text{ gpm})(60 \text{ min/hr}) = 12{,}000 \text{ gph}$$

$$\text{Operating time, hr} = \frac{386{,}400 \text{ gal}}{12{,}000 \text{ gph}} = 32.2 \text{ hr, round to } \mathbf{30\ hr}$$

CALCULATIONS FOR CHEMICAL PRECIPITATION SOFTENING PROCESS

TABLE 3-1 Molecular weights of chemical compounds

Compound	Molecular Weight
Alkalinity, as $CaCO_3$	100.1
Carbon dioxide, CO_2	44.0
Hardness, as $CaCO_3$	100.1
Hydrated lime, $Ca(OH)_2$	74.1
Magnesium, Mg^{2+}	24.3
Magnesium hydroxide, $Mg(OH)_2$	58.3
Quicklime, CaO	56.1
Soda ash, Na_2CO_3	106.0

Use Table 3-1 to solve the problems that follow.

112. Determine the hydrated lime dose required, in milligrams per liter, for water with the following characteristics:

	Source Water	Softened Water Before Blending
Total alkalinity, mg/L	212 mg/L as $CaCO_3$	28 mg/L
Total hardness, mg/L	302 mg/L as $CaCO_3$	41 mg/L
CO_2, mg/L	10.0 mg/L	0 mg/L
Mg^{2+}	27 mg/L	9.0 mg/L
pH	7.5	8.5
Lime purity	90%	

Calculate the hydrated lime required in milligrams per liter. Use an excess lime dosage of 15% (115% or 1.15 in decimal form).

The equation is:

$$\textbf{Hydrated lime feed, mg/L} = \frac{(A + B + C + D)(1.15)}{\textbf{Lime purity}}$$

Where A is CO_2 in source water:

$$A = (\text{mg/L as } CO_2)(74.1/44)$$

where 74.1 = molecular weight (MW) of $Ca(OH)_2$.

Substitute known values and solve.

$$A = (10.0 \text{ mg/L})(74.1/44) = 16.84 \text{ mg/L}$$

Where B is bicarbonate (total) alkalinity removed in softening:

$$B = (\text{mg/L as } CaCO_3 \text{ removed})(74.1/100.1)$$

where 100.1 = MW as $CaCO_3$.

Substitute known values and solve.

$$B = (212 \text{ mg/L} - 28 \text{ mg/L})(74.1/100.1) = 136.2 \text{ mg/L}$$

Where C is hydroxide alkalinity in softened effluent:

$$C = (\text{mg/L as } CaCO_3)(74.1/100.1)$$

In this case, there is no hydroxide alkalinity, therefore, C = 0.

Where D is magnesium removed in softening:

$$D = (\text{mg/L as } Mg^{2+})(74.1/24.3)$$

where 24.3 = MW of Mg^{2+}.

$$D = (27 \text{ mg/L} - 9.0 \text{ mg/L})(74.1/24.3) = 54.89 \text{ mg/L}$$

$$\text{Hydrated lime feed, mg/L} = \frac{(16.84 \text{ mg/L} + 136.2 \text{ mg/L} + 0 + 54.89 \text{ mg/L})(1.15)}{90\%/100\%}$$

Hydrated lime feed, mg/L = 265.69 mg/L, round to **270 mg/L $Ca(OH)_2$**

113. Determine the hydrated lime dose required, in milligrams per liter, for water with the following characteristics:

	Source Water	Softened Water Before Blending
Total alkalinity, mg/L	224 mg/L as $CaCO_3$	33 mg/L
Total hardness, mg/L	309 mg/L as $CaCO_3$	42 mg/L
CO_2, mg/L	9.0 mg/L	0 mg/L
Mg^{2+}	25 mg/L	9.0 mg/L
pH	7.4	8.4
Lime purity	90.0%	

Calculate the hydrated lime required in milligrams per liter. Use an excess lime dosage of 15% (115% or 1.15 in decimal form).

The equation is:

$$\textbf{Hydrated lime feed, mg/L} = \frac{(A + B + C + D)(1.15)}{\textbf{Lime purity}}$$

Where A is CO_2 in source water:

$$A = (\text{mg/L as } CO_2)(74.1/44)$$

where 74.1 = MW of $Ca(OH)_2$.

(continued)

Substitute known values and solve.

$$A = (9.0 \text{ mg/L})(74.1/44) = 15.16 \text{ mg/L}$$

Where B is bicarbonate (total) alkalinity removed in softening:

$$B = (\text{mg/L as } CaCO_3 \text{ removed})(74.1/100.1)$$

where 100.1 = MW as $CaCO_3$.

Substitute known values and solve.

$$B = (224 \text{ mg/L} - 33 \text{ mg/L})(74.1/100.1) = 141.39 \text{ mg/L}$$

Where C is hydroxide alkalinity in softened effluent:

$$C = (\text{mg/L as } CaCO_3)(74.1/100.1)$$

In this case, there is no hydroxide alkalinity, therefore, $C = 0$.

Where D is magnesium removed in softening:

$$D = (\text{mg/L as } Mg^{2+})(74.1/24.3)$$

where 24.3 = MW of Mg^{2+}.

$$D = (25 \text{ mg/L} - 9.0 \text{ mg/L})(74.1/24.3) = 48.79 \text{ mg/L}$$

$$\text{Hydrated lime feed, mg/L} = \frac{(15.16 \text{ mg/L} + 141.39 \text{ mg/L} + 0 + 48.79 \text{ mg/L})(1.15)}{90.0\%/100\%}$$

Hydrated lime feed, mg/L = 262.38 mg/L, round to **260 mg/L** $\mathbf{Ca(OH)_2}$

114. Determine the hydrated lime dose required, in milligrams per liter, for water with the following characteristics:

	Source Water	Softened Water
Total alkalinity, mg/L	225 mg/L as $CaCO_3$	43 mg/L
Total hardness, mg/L	326 mg/L as $CaCO_3$	80 mg/L
CO_2, mg/L	16 mg/L	0 mg/L
Mg^{2+}	22 mg/L	8.0 mg/L
pH	7.0	8.1
Lime purity	92%	

Calculate the hydrated lime required in milligrams per liter. Use an excess lime dosage of 15% (115% or 1.15 in decimal form).

The equation is:

$$\textbf{Hydrated lime feed, mg/L} = \frac{(A + B + C + D)(1.15)}{\textbf{Lime purity}}$$

Where A is CO_2 in source water:

$$A = (\text{mg/L as } CO_2)(74.1/44)$$

where 74.1 = MW of $Ca(OH)_2$.

Substitute known values and solve.

$$A = (16 \text{ mg/L})(74.1/44) = 26.95 \text{ mg/L}$$

Where B is bicarbonate (total) alkalinity removed in softening:

$$B = (\text{mg/L as } CaCO_3 \text{ removed})(74.1/100.1)$$

where 100.1 = MW as $CaCO_3$.

$$B = (225 \text{ mg/L} - 43 \text{ mg/L})(74.1/100.1) = 134.73 \text{ mg/L}$$

Where C is hydroxide alkalinity in softened effluent:

$$C = (\text{mg/L as } CaCO_3)(74.1/100.1)$$

(continued)

In this case, there is no hydroxide alkalinity, therefore, $C = 0$.

Where D is magnesium removed in softening:

$$D = (\text{mg/L as } Mg^{2+})(74.1/24.3)$$

where 24.3 = MW of Mg^{2+}.

$$D = (22 \text{ mg/L} - 8.0 \text{ mg/L})(74.1/24.3) = 42.69 \text{ mg/L}$$

$$\text{Hydrated lime feed, mg/L} = \frac{(26.95 \text{ mg/L} + 134.73 \text{ mg/L} + 0 + 42.69 \text{ mg/L})(1.15)}{92\%/100\%}$$

Hydrated lime feed, mg/L = 255.46 mg/L, round to **260 mg/L $Ca(OH)_2$**

115. Determine the hydrated lime dose required, in milligrams per liter, for water with the following characteristics:

	Source Water	Softened Water Before Blending
Total alkalinity, mg/L	178 mg/L as $CaCO_3$	36 mg/L
Total hardness, mg/L	256 mg/L as $CaCO_3$	52 mg/L
CO_2, mg/L	12 mg/L	0 mg/L
Mg^{2+}	21 mg/L	8.0 mg/L
pH	6.9	8.2
Lime purity	88%	

Calculate the hydrated lime required in milligrams per liter. Use an excess lime dosage of 15% (115% or 1.15 in decimal form).

The equation is:

$$\textbf{Hydrated lime feed, mg/L} = \frac{(A + B + C + D)(1.15)}{\textbf{Lime purity}}$$

Where A is CO_2 in source water:

$$A = (\text{mg/L as } CO_2)(74.1/44)$$

where 74.1 = MW of $Ca(OH)_2$.

Substitute known values and solve.

$$A = (12\text{ mg/L})(74.1/44) = 20.21\text{ mg/L}$$

Where B is bicarbonate (total) alkalinity removed in softening:

$$B = (\text{mg/L as } CaCO_3 \text{ removed})(74.1/100.1)$$

where 100.1 = MW as $CaCO_3$.

Substitute known values and solve.

$$B = (178\text{ mg/L} - 36\text{ mg/L})(74.1/100.1) = 105.12\text{ mg/L}$$

Where C is hydroxide alkalinity in softened effluent:

$$C = (\text{mg/L as } CaCO_3)(74.1/100.1)$$

In this case, there is no hydroxide alkalinity, therefore, $C = 0$.

Where D is magnesium removed in softening:

$$D = (\text{mg/L as } Mg^{2+})(74.1/24.3)$$

where 24.3 = MW of Mg^{2+}.

$$D = (21\text{ mg/L} - 8.0\text{ mg/L})(74.1/24.3) = 39.64\text{ mg/L}$$

$$\text{Hydrated lime feed, mg/L} = \frac{(20.21\text{ mg/L} + 105.12\text{ mg/L} + 0 + 39.64\text{ mg/L})(1.15)}{88\%/100\%}$$

Hydrated lime feed, mg/L = 215.59 mg/L, round to **220 mg/L $Ca(OH)_2$**

116. Determine the hydrated lime dose required, in milligrams per liter, for water with the following characteristics:

	Source Water	Softened Water
Total alkalinity, mg/L	280 mg/L as $CaCO_3$	68 mg/L
Total hardness, mg/L	415 mg/L as $CaCO_3$	97 mg/L
CO_2, mg/L	14 mg/L	0 mg/L
Mg^{2+}	36 mg/L	10.0 mg/L
pH	7.6	8.8
Lime purity	90.0%	

Calculate the hydrated lime required in milligrams per liter. Use an excess lime dosage of 15% (115% or 1.15 in decimal form).

The equation is:

$$\textbf{Hydrated lime feed, mg/L} = \frac{(A + B + C + D)(1.15)}{\textbf{Lime purity}}$$

Where A is CO_2 in source water:

$$A = (\text{mg/L as } CO_2)(74.1/44)$$

where 74.1 = MW of $Ca(OH)_2$.

Substitute known values and solve.

$$A = (14 \text{ mg/L})(74.1/44) = 23.58 \text{ mg/L}$$

Where B is bicarbonate (total) alkalinity removed in softening:

$$B = (\text{mg/L as } CaCO_3 \text{ removed})(74.1/100.1)$$

where 100.1 = MW as $CaCO_3$.

$$B = (280 \text{ mg/L} - 68 \text{ mg/L})(74.1/100.1) = 156.94 \text{ mg/L}$$

Where C is hydroxide alkalinity in softened effluent:

$$C = (\text{mg/L as } CaCO_3)(74.1/100.1)$$

In this case, there is no hydroxide alkalinity, therefore, $C = 0$.

Where D is magnesium removed in softening:

$$D = (\text{mg/L as } Mg^{2+})(74.1/24.3)$$

where 24.3 = MW of Mg^{2+}.

$$D = (36 \text{ mg/L} - 10.0 \text{ mg/L})(74.1/24.3) = 79.28 \text{ mg/L}$$

$$\text{Hydrated lime feed, mg/L} = \frac{(23.58 \text{ mg/L} + 156.94 \text{ mg/L} + 0 + 79.28 \text{ mg/L})(1.15)}{90.0\%/100\%}$$

Hydrated lime feed, mg/L = 331.97 mg/L, round to **330 mg/L $Ca(OH)_2$**

POUNDS REMOVAL PROBLEMS

117. Find the amount of algae removed per year if the algae concentration is 0.80 mg/L and the plant treats an average of 6.7 mgd. The removal efficiency is 97% (0.97).

The equation is:

Number of lb/yr

= (mgd)(365 d/yr)(8.34 lb/gal)(Dosage, mg/L)(% Removal efficiency)

Pounds per year algae removal = (6.7 mgd)(365 d/yr)(8.34 lb/gal)(0.80 mg/L)(0.97)

Pounds per year algae removal

= 15,827 lb/yr, round to **16,000 lb/yr of algae removed**

118. Find the amount of iron (Fe) removed per year from a plant that treats an average of 35 mgd if the average Fe concentration is 0.6 ppm and the removal efficiency is 83%.

First, calculate the Fe removal in parts per million.

$$(0.6 \text{ ppm})(83\%/100\%) = 0.6 \text{ ppm}(0.83) = 0.498 \text{ ppm} = 0.498 \text{ mg/L}$$

Determine the amount of water in million gallons produced for the year.

$$(35 \text{ mgd})(365 \text{ d/yr}) = 12{,}775 \text{ mil gal/yr}$$

Next, using the "pounds" equation, solve for the number of pounds per year.

$$\textbf{Pounds per year} = \textbf{(mil gal/yr)(Dosage, mg/L)(8.34 lb/gal)}$$

$$\text{Pounds per year, Fe removal} = (12{,}775 \text{ mil gal/yr})(0.498 \text{ mg/L})(8.34 \text{ lb/gal})$$

= 53,059 lb/yr, round to **50,000 lb/yr, Fe removed**

119. Find the amount of manganese (Mn) removed per year from a plant that treats an average of 38 mgd if the average Mn concentration is 0.10 ppm and the removal efficiency is 86%.

First, calculate the Mn removal in parts per million.

$$(0.10 \text{ ppm})(86\%/100\%) = 0.10 \text{ ppm}(0.86) = 0.086 \text{ ppm (i.e., 0.086 mg/L)}$$

Determine the amount of water in million gallons produced for the year.

$$(38 \text{ mgd})(365 \text{ d/yr}) = 13{,}870 \text{ mil gal/yr}$$

Next, using the "pounds" equation, solve for the number of pounds per year.

$$\textbf{Pounds per year} = \textbf{(mil gal/yr)(Dosage, mg/L)(8.34 lb/gal)}$$

$$\text{Pounds per year, Mn removal} = (13{,}870 \text{ mil gal/yr})(0.086 \text{ mg/L})(8.34 \text{ lb/gal})$$

= 9,948.12 lb/yr, round to **9,900 lb/yr, Mn removed**

120. Find the amount of algae removed per year if the algae concentration is 0.60 mg/L and the plant treats an average of 11.8 mgd. The removal efficiency is 96% (0.96).

The equation is:

Pounds per year

= (mgd)(365 d/yr)(8.34 lb/gal)(Dosage, mg/L)(% Removal efficiency)

Pounds per year, algae removal = (11.8 mgd)(365 d/yr)(8.34 lb/gal)(0.60 mg/L)(0.96)

Pounds per year, algae removal = 20,690 lb/yr, round to **21,000 lb/yr of algae removed**

121. Find the amount of zinc (Zn) removed per year from a plant that treats an average of 12.5 mgd if the average Zn concentration is 0.012 ppm and the removal efficiency is 83%.

First, calculate the Zn removal in parts per million.

(0.012 ppm)(83%/100%) = (0.012 ppm)(0.83) = 0.00996 ppm (i.e., 0.0996 mg/L)

Determine the amount of water in million gallons produced for the year.

(12.5 mgd)(365 d/yr) = 4,562.5 mil gal/yr

Next, using the "pounds" equation, solve for the number of pounds per year.

Pounds per year = (mil gal/yr)(Dosage, mg/L)(8.34 lb/gal)

Pounds per year, Zn removal = (4,562.5 mil gal/yr)(0.00996 mg/L)(8.34 lb/gal)

= 378.99 lb/yr, round to **380 lb/yr, Zn removed**

122. Find the amount of zinc (Zn) removed per year from a plant that treats an average of 62 mgd if the average Zn concentration is 0.0110 ppm and the removal efficiency is 78%.

First, calculate the Zn removal in parts per million.

(0.0110 ppm)(78%/100%) = 0.0110 ppm(0.78) = 0.00858 ppm (i.e., 0.00858 mg/L)

Determine the amount of water in million gallons produced for the year.

(62 mgd)(365 d/yr) = 22,630 mil gal/yr

Next, using the "pounds" equation, solve for the number of pounds per year.

Pounds per year = (mil gal/yr)(Dosage, mg/L)(8.34 lb/gal)

Pounds per year, Zn removal = (22,630 mil gal/yr)(0.00858 mg/L)(8.34 lb/gal)

= 1,619.34 lb/yr, round to **1,600 lb/yr, Zn removed**

123. Find the amount of iron (Fe) removed per year from a plant that treats an average of 18.5 mgd if the average Fe concentration is 0.31 mg/L and the removal efficiency is 86%.

First, calculate the Fe removal in parts per million.

(0.31 mg/L)(86%/100%) = 0.31 mg/L(0.86) = 0.267 mg/L

Determine the amount of water in million gallons produced for the year.

(18.5 mgd)(365 d/yr) = 6,752.5 mil gal/yr

Next, using the "pounds" equation, solve for the number of pounds per year.

Pounds per year = (mil gal/yr)(Dosage, mg/L)(8.34 lb/gal)

Pounds per year, Fe removal = (6,752.5 mil gal/yr)(0.267 mg/L)(8.34 lb/gal)

= 15,036 lb/yr, round to **15,000 lb/yr, Fe removed**

124. Determine the pounds per month (30 days) of algae removed if the flow is 21 mgd, the average algae concentration is 0.35 mg/L, and the removal efficiency is 96%.

First, determine the amount of algae removed per day using the "pounds" formula.

Pounds per day, algae removed per day

$$= \textbf{(mgd)(Concentration, mg/L)(8.34 lb/gal)(Efficiency of removal)}$$

Algae removed per day = (21 mgd)(0.35 mg/L)(8.34 lb/gal)(96%/100%) = 58.85 lb/d

Next, find the pounds of algae removed in one month.

(58.85 lb/d)(30 d/month)

= 1,765.5 lb/month, round to **1,800 lb/month of algae removed**

FLOW RATE CALCULATIONS

125. A channel (rectangular shape) is 6.0 ft wide with water flowing through it at a velocity of 2.3 ft/s. What is the depth of the channel if the flow through the channel is 22.7 ft^3/s?

The equation is:

$$\textbf{Flow, ft}^3\textbf{/s} = \textbf{(Area)(Velocity, ft/s)}$$

Where Area = (Width)(Depth). Depth is unknown.

$$22.7 \text{ ft}^3/\text{s} = (6.0 \text{ ft})(\text{Depth, ft})(2.3 \text{ ft/s})$$

Solve for the depth by rearranging the equation and substituting the known values.

$$\text{Depth, ft} = \frac{22.7 \text{ ft}^3/\text{s}}{(6.0 \text{ ft})(2.3 \text{ ft/s})} = 1.645 \text{ ft, round to } \textbf{depth of 1.6 ft}$$

126. Water is flowing at a velocity of 3.25 ft/s in an 8.0-in. diameter pipe. If the pipe changes from an 8.0-in. pipe to a 12-in. pipe, what will the velocity be in the 12-in. pipe?

Flow in the 8.0-in. pipe equals flow in the 12-in. pipe because the flow must remain constant.

$$\mathbf{Q_1 = Q_2}$$

Q, Flow = (Area)(Velocity), it follows that:

$$\textbf{(Area 1)(Velocity 1)} = \textbf{(Area 2)(Velocity 2)}$$

First, find the diameters, in feet, for the 8.0-in. and 12-in. pipes.

$$\text{Diameter for 8.0-in.} = \text{8.0-in.}(1\text{ ft}/12\text{ in.}) = 0.667\text{ ft}$$

$$\text{Diameter for 12-in.} = \text{12-in.}(1\text{ ft}/12\text{ in.}) = 1.0\text{ ft}$$

Then, determine the areas of each size pipe.

$$\textbf{Area} = \textbf{(0.785)(Diameter)}^2$$

$$\text{Area 1 (8.0-in.)} = (0.785)(0.667\text{ ft})(0.667\text{ ft}) = 0.349\text{ ft}^2$$

$$\text{Area 2 (12-in.)} = (0.785)(1.0\text{ ft})(1.0\text{ ft}) = 0.785\text{ ft}^2$$

Last, substitute areas calculated and known velocity in 8.0-in. pipe.

$$(0.349\text{ ft}^2)(3.25\text{ ft/s}) = (0.785\text{ ft}^2)(x\text{, ft/s})$$

Solve for x.

$$x\text{, ft/s} = \frac{(0.349\text{ ft}^2)(3.25\text{ ft/s})}{(0.785\text{ ft}^2)} = 1.445\text{ ft/s, round to } \textbf{1.4 ft/s in the 12-in. pipe}$$

127. Water is flowing at a velocity of 1.6 ft/s in a 4-in. diameter pipe. If the pipe changes from the 4-in. pipe to a 3-in. pipe, what will the velocity be in the 3-in. pipe?

Flow in the 4-in. pipe equals flow in the 3-in. pipe because the flow must remain constant.

$$\mathbf{Q_1 = Q_2}$$

The equation is:

$$\mathbf{(Area\ 1)(Velocity\ 1) = (Area\ 2)(Velocity\ 2)}$$

First, find the diameter for the 3-in. and 4-in. pipes.

$$\text{Diameter for 3-in.} = (\text{3-in.})(1\ \text{ft}/12\ \text{in.}) = 0.25\ \text{ft}$$

$$\text{Diameter for 4-in.} = (\text{4-in.})(1\ \text{ft}/12\ \text{in.}) = 0.333\ \text{ft}$$

Then, determine the areas of each size pipe.

$$\mathbf{Area = (0.785)(Diameter)^2}$$

$$\text{Area 1 (3-in.)} = (0.785)(0.25\ \text{ft})(0.25\ \text{ft}) = 0.049\ \text{ft}^2$$

$$\text{Area 2 (4-in.)} = (0.785)(0.333\ \text{ft})(0.333\ \text{ft}) = 0.087\ \text{ft}^2$$

Last, substitute areas calculated and known velocity in 4-in. pipe.

$$(0.049\ \text{ft}^2)(x,\ \text{ft/s}) = (0.087\ \text{ft}^2)(1.6\ \text{ft/s})$$

Solve for x.

$$x,\ \text{ft/s} = \frac{(0.087\ \text{ft}^2)(1.6\ \text{ft/s})}{(0.049\ \text{ft}^2)} = 2.84\ \text{ft/s, round to } \mathbf{3\ ft/s\ in\ the\ 3\text{-}in.\ pipe}$$

128. Water is flowing at a velocity of 4.2 ft/s in an 8.0-in. diameter pipe. If the pipe changes from the 8.0-in. pipe to a 10-in. pipe, what will the velocity be in the 10-in. pipe?

Flow in the 8.0-in. pipe equals flow in the 10-in. pipe because the flow must remain constant.

$$\mathbf{Q_1 = Q_2}$$

Q, Flow = (Area)(Velocity), it follows that:

$$\textbf{(Area 1)(Velocity 1)} = \textbf{(Area 2)(Velocity 2)}$$

First, find the diameters, in feet, for the 8-in. and 10-in. pipes.

$$\text{Diameter for 8.0-in.} = \text{8.0-in.}(1 \text{ ft}/12 \text{ in.}) = 0.667 \text{ ft}$$

$$\text{Diameter for 10-in.} = \text{10-in.}(1 \text{ ft}/12 \text{ in.}) = 0.833 \text{ ft}$$

Then, determine the area of each size pipe.

$$\textbf{Area} = \textbf{(0.785)(Diameter)}^2$$

$$\text{Area 1 (8.0-in.)} = (0.785)(0.667 \text{ ft})(0.667 \text{ ft}) = 0.349 \text{ ft}^2$$

$$\text{Area 2 (10-in.)} = (0.785)(0.833 \text{ ft})(0.833 \text{ ft}) = 0.545 \text{ ft}^2$$

Last, substitute areas calculated and known velocity in 8.0-in. pipe.

$$(0.349 \text{ ft}^2)(4.2 \text{ ft/s}) = (0.545 \text{ ft}^2)(x, \text{ ft/s})$$

Solve for x.

$$x, \text{ ft/s} = \frac{(0.349 \text{ ft}^2)(4.2 \text{ ft/s})}{(0.545 \text{ ft}^2)} = 2.69 \text{ ft/s, round to } \textbf{2.7 ft/s in the 10-in. pipe}$$

129. Water is flowing at a velocity of 0.85 ft/s in a 12-in. diameter pipe. If the pipe changes from the 12-in. pipe to an 8.0-in. pipe, what will the velocity be in the 8-in. pipe?

Flow in the 12-in. pipe equals flow in the 8.0-in. pipe because the flow must remain constant.

$$\mathbf{Q_1 = Q_2}$$

The equation is:

$$\textbf{(Area 1)(Velocity 1)} = \textbf{(Area 2)(Velocity 2)}$$

First, find the diameter for the 8.0-in. and 12-in. pipes.

$$\text{Diameter for 8.0-in.} = (\text{8.0-in.})(1\text{ ft}/12\text{ in.}) = 0.667\text{ ft}$$

$$\text{Diameter for 12-in.} = (\text{12-in.})(1\text{ ft}/12\text{ in.}) = 1.0\text{ ft}$$

Then, determine the area of each size pipe.

$$\textbf{Area} = \textbf{(0.785)(Diameter)}^{\mathbf{2}}$$

$$\text{Area 1 (8.0-in.)} = (0.785)(0.667\text{ ft})(0.667\text{ ft}) = 0.349\text{ ft}^2$$

$$\text{Area 2 (12-in.)} = (0.785)(1.0\text{ ft})(1.0\text{ ft}) = 0.785\text{ ft}^2$$

Last, substitute areas calculated and known velocity in 12-in. pipe.

$$(0.349\text{ ft}^2)(x\text{, ft/s}) = (0.785\text{ ft}^2)(0.85\text{ ft/s})$$

Solve for x.

$$x\text{, ft/s} = \frac{(0.785\text{ ft}^2)(0.85\text{ ft/s})}{(0.349\text{ ft}^2)} = 1.91\text{ ft/s, round to } \textbf{1.9 ft/s in the 8.0-in. pipe}$$

130. Water is flowing at a velocity of 1.25 ft/s in a 10.0-in. diameter pipe. If the pipe changes from the 10.0-in. pipe to a 6.0-in. pipe, what will the velocity be in the 6-in. pipe?

Flow in the 10.0-in. pipe equals flow in the 6.0-in. pipe because the flow must remain constant.

$$\mathbf{Q_1 = Q_2}$$

The equation is:

$$\textbf{(Area 1)(Velocity 1)} = \textbf{(Area 2)(Velocity 2)}$$

First, find the diameter for the 6.0-in. and 10.0-in. pipes.

$$\text{Diameter for 6.0-in.} = (\text{6.0-in.})(1\text{ ft}/12\text{ in.}) = 0.5\text{ ft}$$

$$\text{Diameter for 10.0-in.} = (\text{10.0-in.})(1\text{ ft}/12\text{ in.}) = 0.833\text{ ft}$$

Then, determine the area of each size pipe.

$$\textbf{Area} = \mathbf{(0.785)(Diameter)^2}$$

$$\text{Area 1 (6.0-in.)} = (0.785)(0.5\text{ ft})(0.5\text{ ft}) = 0.196\text{ ft}^2$$

$$\text{Area 2 (10.0-in.)} = (0.785)(0.833\text{ ft})(0.833\text{ ft}) = 0.545\text{ ft}^2$$

Last, substitute areas calculated and known velocity in 10-in. pipe.

$$(0.196\text{ ft}^2)(x, \text{ft/s}) = (0.545\text{ ft}^2)(1.25\text{ ft/s})$$

Solve for x.

$$x, \text{ft/s} = \frac{(0.545\text{ ft}^2)(1.25\text{ ft/s})}{(0.196\text{ ft}^2)} = 3.476\text{ ft/s, round to } \textbf{3.5 ft/s in the 6.0-in. pipe}$$

PUMPING RATE CALCULATIONS

131. The level in a storage tank drops 4.4 ft in 5.0 hr. If the tank has a diameter of 100.0 ft and the plant is producing 5.6 mgd, what is the average discharge rate of the treated water discharge pumps in gallons per minute?

First, find the water production during the 5.0-hr interval.

$$\text{Water produced in 5.0-hr interval} = \frac{(\text{mgd})(1{,}000{,}000)(1\text{ d})(5.0\text{ hr})}{(1\text{ M})(24\text{ hr})}$$

$$\text{Water produced in 5.0-hr interval} = \frac{(5.6\text{ mgd})(1{,}000{,}000)(1\text{ d})(5.0\text{ hr})}{(1\text{ M})(24\text{ hr})}$$

$$= 1{,}166{,}667\text{ gal}$$

Next, find the gallons contained in the 4.4-ft drop in water level.

$$\textbf{Volume, tank} = \textbf{(0.785)(Diameter)}^2\textbf{(Drop in height)}$$

Volume of 4.4 ft in 100.0 ft diameter tank

$$= (0.785)(100.0\text{ ft})(100.0\text{ ft})(4.4\text{ ft})(7.48\text{ gal/ft}^3)$$

Volume of 4.4 ft in 100.0 ft diameter tank = 258,359 gal

Production plus the loss in level is the amount the discharge pumps had to send into the distribution system, but first find the number of minutes in 5.0 hr.

$$(5.0\text{ hr})(60\text{ min/hr}) = 300\text{ min}$$

Then, determine total gallons the discharge pumps moved.

$$\text{Total gallons discharge pumps moved in 5.0 hr} = 1{,}166{,}667\text{ gal} + 258{,}359\text{ gal}$$

$$= 1{,}425{,}026\text{ gal}$$

Last, divide the number of gallons the discharge pumps moved by the time in minutes.

Discharge pumps, gpm = 1,425,026 gal/300 min = 4,750 gpm, round to **4,800 gpm**

132. The level in a storage tank drops 7.9 ft in exactly 8 hr. If the tank has a diameter of 60 ft and the plant is producing 3.8 mgd, what is the average discharge rate of the treated water discharge pump in gallons per minute?

First, find the water production during the 8-hr interval.

Gallons of water treated in 8-hr interval = (3.8 mgd)(1,000,000)(8 hr)/24 hr

= 1,266,667 gal

Next, find the gallons contained in the 7.9-ft drop in water level.

Volume, tank = (0.785)(Diameter)^2(Drop in height)

Volume of 7.9 ft in 60 ft diameter tank = (0.785)(60 ft)(60 ft)(7.9 ft)(7.48 gal/ft^3)

= 166,994 gal

Production plus the loss in level is the amount the discharge pump had to send into the distribution system, but first find the number of minutes in 8 hr.

(8 hr)(60 min/hr) = 480 min

Then, determine total gallons the discharge pump moved.

Total gallons discharge pump moved in 8 hr = 1,266,667 gal + 166,994 gal

= 1,433,661 gal

Last, divide the number of gallons the discharge pump moved by the time in minutes.

Discharge pump, gpm = 1,433,661 gal/480 min = 2,987 gpm, round to **3,000 gpm**

133. The level in a clear well raises 1.35 ft in exactly 2 hr. If the clear well has a length of 375 ft, a width of 80.0 ft, and the plant is producing 65 mgd, what is the average discharge rate of the treated water discharge pumps in gallons per minute?

First, find the water production during the 2-hr interval.

Gallons of water treated in 2 hr = (65 mgd)(1,000,000/M)(2 hr)/24 hr = 5,416,667 gal

Next, find the gallons contained in the clear well's rise of 1.35 ft.

Volume of clear well = (Length)(Width)(Height of increase)

Volume of clear well = (375 ft)(80.0 ft)(1.35 ft)(7.48 gal/ft^3) = 302,940 gal

Production minus the rise in the water level is the amount the discharge pumps had to send into the distribution system, but first find the number of minutes in 2 hr.

(2 hr)(60 min/hr) = 120 min

Then, determine total gallons the discharge pumps moved.

Total gallons discharge pumps moved in 2 hr = 5,416,667 gal – 302,940 gal

= 5,113,727 gal

Last, divide the number of gallons the discharge pumps moved by the time in minutes.

Discharge pumps, gpm = 5,113,727 gal/120 min = 42,614 gpm, round to **43,000 gpm**

134. The level in a clear well falls 2.76 ft in 2.2 hr. If the clear well has a length of 245 ft, a width of 85 ft, and the plant is producing 14.3 mgd, what is the average discharge rate of the treated water discharge pumps in gallons per minute?

First, find the water production during the 2.2-hr interval.

Gallons of water treated in 2.2 hr = (14.3 mgd)(1,000,000/M)(2.2 hr)/24 hr

= 1,310,833 gal

Next, find the gallons contained in the clear well's decrease of 2.76 ft.

Volume of clear well decrease = (Length)(Width)(Height of decrease)

Volume of clear well = (245 ft)(85 ft)(2.76 ft)(7.48 gal/ft^3) = 429,928 gal

Production plus the fall in the water level is the amount the discharge pumps had to send into the distribution system, but first find the number of minutes in 2.2 hr.

(2.2 hr)(60 min/hr) = 132 min

Then, determine total gallons the discharge pumps moved.

Total gallons discharge pumps moved in 2.2 hr = 1,310,833 gal + 429,928 gal

= 1,740,761 gal

Last, divide the number of gallons the discharge pumps moved by the time in minutes.

Discharge pumps, gpm = 1,740,761 gal/132 min = 13,188 gpm, round to **13,000 gpm**

135. The level in a storage tank drops 2.5 ft in 3.0 hr. If the tank has a diameter of 120 ft and the plant is producing 7.1 mgd, what is the average discharge rate of the treated water discharge pumps in gallons per minute?

First, find the water production during the 3-hr interval.

$$\text{Water produced in 3.0-hr interval} = \frac{(\text{mgd})(1{,}000{,}000)(1\text{ d})(3\text{ hr})}{(1\text{ M})(24\text{ hr})}$$

Substituting:

$$(7.1\text{ mgd})(1{,}000{,}000)(3.0\text{ hr})/(24\text{ hr/d}) = 887{,}500\text{ gal}$$

Next, find the gallons contained in the 2.5-ft drop in water level.

$$\textbf{Volume, tank} = \textbf{(0.785)(Diameter)}^2\textbf{(Drop in height)}$$

$$\text{Volume of 2.5 ft in 100-ft diameter tank} = (0.785)(120\text{ ft})(120\text{ ft})(2.5\text{ ft})(7.48\text{ gal/ft}^3)$$

$$= 211{,}385\text{ gal}$$

Production plus the loss in level is the amount the discharge pumps had to send into the distribution system, but first find the number of minutes in 3 hr.

$$3.0\text{ hr}(60\text{ min/hr}) = 180\text{ min}$$

Then, determine total gallons the discharge pumps moved.

$$\text{Total gallons discharge pumps moved in 3.0 hr} = 887{,}500\text{ gal} + 211{,}385\text{ gal}$$

$$= 1{,}098{,}885\text{ gal}$$

Last, divide the number of gallons the discharge pumps moved by the time in minutes.

Discharge pumps, gpm = 1,098,885 gal/180 min = 6,105 gpm, round to **6,100 gpm**

136. The level in a clear well falls 3.05 ft in 3.5 hr. If the clear well has a length of 190 ft, a width of 80.25 ft, and the plant is producing 9.13 mgd, what is the average discharge rate of the treated water discharge pumps in gallons per minute?

First, find the water production during the 3.5-hr interval.

$$\text{Gallons of water treated in 3.5-hr interval} = (9.13\text{ mgd})(1{,}000{,}000)(3.5\text{ hr})/24\text{ hr}$$

$$= 1{,}331{,}458\text{ gal}$$

Next, find the gallons contained in the clear well's decrease of 3.05 ft.

$$\textbf{Volume of clear well decrease} = \textbf{(Length)(Width)(Drop in height)}$$

$$\text{Volume of clear well} = (190\text{ ft})(80.25\text{ ft})(3.05\text{ ft})(7.48\text{ gal/ft}^3) = 347{,}856\text{ gal}$$

Production plus the fall in the water level is the amount the discharge pumps had to send into the distribution system, but first find the number of minutes in 3.5 hr.

$$(3.5 \text{ hr})(60 \text{ min/hr}) = 210 \text{ min}$$

Then, determine total gallons the discharge pumps moved.

$$\begin{aligned}\text{Total gallons discharge pumps moved in 3.5 hr} &= 1{,}331{,}458 \text{ gal} + 347{,}856 \text{ gal}\\ &= 1{,}679{,}314 \text{ gal}\end{aligned}$$

Last, divide the number of gallons the discharge pumps moved by the time in minutes.

Discharge pumps, gpm = 1,679,314 gal/210 min = 7,997 gpm, round to **8,000 gpm**

PUMPING PROBLEMS

137. The difference between the inlet and outlet pressure gauges for a pump that is off is 72 psi. What is the total head if friction and minor head losses are 11 ft?

First, find the static head by converting the number of pounds per square inch to feet using conversion of 2.31 ft/psi.

$$\text{Static head, ft} = (72 \text{ psi})(2.31 \text{ ft/psi}) = 166.32 \text{ ft}$$

The equation for total head is:

$$\textbf{Total head, ft} = \textbf{Static head, ft} + \textbf{Head losses, ft}$$

Total head, ft = 166.32 ft + 11 ft = 177.32 ft, round to **180 ft**

138. What is the total head, in feet, for a pump that is operating when the inlet pressure gauge reads 79 psi and the outlet gauge reads 128 psi?

For a pump in operation:

$$\text{Total head, ft} = \text{Static head, ft}$$

The static head is the difference in the pressure gauges.

$$\begin{aligned}\text{Static head, ft} &= (128 \text{ psi} - 79 \text{ psi})(2.31 \text{ ft/psi})\\ &= (49 \text{ psi})(2.31 \text{ ft/psi})\\ &= 113.19 \text{ ft, round to } \mathbf{110 \text{ ft}}\end{aligned}$$

139. What is the total head for a pump that is operating when the inlet pressure gauge reads 62 psi and the outlet gauge reads 114 psi?

For a pump in operation:

$$\text{Total head, ft} = \text{Static head, ft}$$

The static head is the difference in the pressure gauges.

$$\text{Static head, ft} = (114 \text{ psi} - 62 \text{ psi})(2.31 \text{ ft/psi})$$

$$\text{Static head, ft} = (52 \text{ psi})(2.31 \text{ ft/psi}) = 120.12 \text{ ft, round to } \mathbf{120\ ft}$$

140. What is the total head for a pump that is operating when the inlet pressure gauge reads 65 psi and the outlet gauge reads 100 psi?

For a pump in operation:

$$\text{Total head, ft} = \text{Static head, ft}$$

The static head is the difference in the pressure gauges.

$$\text{Static head, ft} = (100 \text{ psi} - 65 \text{ psi})(2.31 \text{ ft/psi})$$

$$\text{Static head, ft} = (35 \text{ psi})(2.31 \text{ ft/psi}) = 80.85 \text{ ft, round to } \mathbf{81\ ft}$$

141. Determine the cost to operate a 100-hp motor for 1 month (assume 30 days) if it runs an average of 7.23 hr/d, is 83% efficient, and the electrical costs are $0.045 per kW.

The equation is:

$$\mathbf{Cost,\ month} =$$

$$\mathbf{(hp)(hr/d)(Number\ of\ days)(0.746\ kW/hp)(Cost/kW\text{-}hr)(\%\ Efficiency)}$$

Substitute known values.

$$\text{Cost, month} = (100 \text{ hp})(7.23 \text{ hr/d})(30 \text{ d})(0.746 \text{ kW/hp})(\$0.045/\text{kW})(83\%/100\%)$$

$$\text{Cost, month} = \mathbf{\$604.35}$$

HYDRAULIC PRESS CALCULATIONS

142. A force of 50.0 lb is applied to a small cylinder on a hydraulic jack. The diameter of the small cylinder is 10.0 in. If the diameter of the large cylinder is 2.5 ft, what is the total lifting force?

The equation is:

$$\textbf{Pressure} = \frac{\textbf{Force, lb}}{\textbf{Area, ft}^2} \quad \text{for pressure on the small cylinder.}$$

First, convert 10.0 in. to feet.

$$(10.0 \text{ in.})(1 \text{ ft}/12 \text{ in.}) = 0.833 \text{ ft}$$

Substituting:

$$\text{Pressure} = \frac{50.0 \text{ lb}}{(0.785)(0.833 \text{ ft})(0.833 \text{ ft})} = 91.79 \text{ lb/ft}^2$$

Next, calculate the total force on the large cylinder.

Rearrange the above equation and then solve for total force.

$$\textbf{Total force} = \textbf{(Pressure)(Area)}$$

Total force = $(91.79 \text{ lb/ft}^2)(0.785)(2.5 \text{ ft})(2.5 \text{ ft}) = 450.34$ lb, round to **450 lb**

143. A small cylinder on a hydraulic jack is 8 in. in diameter. A force of 100 lb is applied to the small cylinder. If the diameter of the large cylinder is 3 ft, what is the total lifting force?

The equation is:

$$\textbf{Pressure} = \frac{\textbf{Force, lb}}{\textbf{Area, ft}^2} \quad \text{for pressure on the small cylinder.}$$

First, convert 6 in. to feet.

$$(8 \text{ in.})(1 \text{ ft}/12 \text{ in.}) = 0.667 \text{ ft}$$

$$\text{Pressure} = \frac{100.0 \text{ lb}}{(0.785)(0.667 \text{ ft})(0.667 \text{ ft})} = 286.34 \text{ lb/ft}^2$$

Next, calculate the total force on the large cylinder.

Rearrange the above equation and then solve for total force.

$$\textbf{Total force} = \textbf{(Pressure)(Area)}$$

Total force = $(286.34 \text{ lb/ft}^2)(0.785)(3 \text{ ft})(3 \text{ ft}) = 2{,}022.99$ lb, round to **2,000 lb**

144. A small cylinder on a hydraulic jack is 6.0 in. in diameter. A force of 200.0 lb is applied to the small cylinder. If the diameter of the large cylinder is 3.0 ft, what is the total lifting force?

The equation is:

$$\textbf{Pressure} = \frac{\textbf{Force, lb}}{\textbf{Area, ft}^2} \quad \text{for pressure on the small cylinder.}$$

First, convert 6 in. to feet.

$$(6.0 \text{ in.})(1 \text{ ft}/12 \text{ in.}) = 0.5 \text{ ft}$$

$$\text{Pressure} = \frac{200.0 \text{ lb}}{(0.785)(0.5 \text{ ft})(0.5 \text{ ft})} = 1{,}019 \text{ lb/ft}^2$$

Next, calculate the total force on the large cylinder.

Rearrange the above equation and then solve for total force.

$$\textbf{Total force} = \textbf{(Pressure)(Area)}$$

Total force = (1,019 lb/ft^2)(0.785)(3.0 ft)(3.0 ft) = 7,199 lb, round to **7,200 lb**

145. A force of 10.0 lb is applied to a small cylinder on a hydraulic jack. The diameter of the small cylinder is 3.0 in. If the diameter of the large cylinder is 1.75 ft, what is the total lifting force?

The equation is:

$$\textbf{Pressure} = \frac{\textbf{Force, lb}}{\textbf{Area, ft}^2} \quad \text{for pressure on the small cylinder.}$$

First, convert 3 in. to feet.

$$(3.0 \text{ in.})(1 \text{ ft}/12 \text{ in.}) = 0.25 \text{ ft}$$

$$\text{Pressure} = \frac{10.0 \text{ lb}}{(0.785)(0.25 \text{ ft})(0.25 \text{ ft})} = 203.82 \text{ lb/ft}^2$$

Next, calculate the total force on the large cylinder.

Rearrange the above equation and then solve for total force.

$$\textbf{Total force} = \textbf{(Pressure)(Area)}$$

Total force = (203.82 lb/ft^2)(0.785)(1.75 ft)(1.75 ft) = 489.996 lb, round to **490 lb**

146. A force of 25 lb is applied to a small cylinder on a hydraulic jack. The diameter of the small cylinder is 6.0 in. If the diameter of the large cylinder is 1.5 ft, what is the total lifting force?

The equation is:

$$\textbf{Pressure} = \frac{\textbf{Force, lb}}{\textbf{Area, ft}^2} \quad \text{for pressure on the small cylinder.}$$

First, convert 6.0 in. to feet.

$$(6.0 \text{ in.})(1 \text{ ft}/12 \text{ in.}) = 0.50 \text{ ft}$$

$$\text{Pressure} = \frac{25 \text{ lb}}{(0.785)(0.50 \text{ ft})(0.50 \text{ ft})} = 127.39 \text{ lb/ft}^2$$

Next, calculate the total force on the large cylinder.

Rearrange the above equation and then solve for total force.

$$\textbf{Total force} = \textbf{(Pressure)(Area)}$$

Total force = $(127.39 \text{ lb/ft}^2)(0.785)(1.5 \text{ ft})(1.5 \text{ ft}) = 225$ lb, round to **220 lb**

147. A small cylinder on a hydraulic jack is 6.0 in. in diameter. A force of 133 lb is applied to the small cylinder. If the diameter of the large cylinder is 2.5 ft, what is the total lifting force?

The equation is:

$$\textbf{Pressure} = \frac{\textbf{Force, lb}}{\textbf{Area, ft}^2} \quad \text{for pressure on the small cylinder.}$$

First, convert 6 in. to feet.

$$(6.0 \text{ in.})(1 \text{ ft}/12 \text{ in.}) = 0.50 \text{ ft}$$

$$\text{Pressure} = \frac{133 \text{ lb}}{(0.785)(0.50 \text{ ft})(0.50 \text{ ft})} = 677.7 \text{ lb/ft}^2$$

Next, calculate the total force on the large cylinder.

Rearrange the above equation and then solve for total force.

$$\textbf{Total force} = \textbf{(Pressure)(Area)}$$

Total force = $(677.7 \text{ lb/ft}^2)(0.785)(2.5 \text{ ft})(2.5 \text{ ft}) = 3{,}325$ lb, round to **3,300 lb**

Water Treatment Grade 3 Test

(answers on p. 207)

1. A well is to be disinfected with 65% calcium hypochlorite. The well is 284 ft deep and 1.5 ft in diameter for the first 100 ft and 1.0 ft in diameter for the remaining length. Depth to water from the top of the casing is 74 ft. If the desired dose is 50 mg/L, how many pounds of calcium hypochlorite are required?

2. Determine the ion exchange softener operating time in hours given the following:
   Exchange capacity of the softener is 7,498,000 grains
   Flow rate is 320 gpm
   Raw water contains 18.3 gpg

3. A well that is 306 ft deep and 14 in. in diameter requires disinfection. Depth to water from top of casing is 114 ft. If the desired dose is 50 mg/L, how many pounds of calcium hypochlorite (65% available chlorine) are required?

4. Water is flowing at a velocity of 2.90 ft/s in a 10-in. diameter pipe. If the pipe changes from the 10-in. pipe to a 12-in. pipe, what will the velocity be in the 12-in. pipe?

5. How many gallons of a 12.8% solution must be mixed with a 4.1% solution to make exactly 325 gal of an 8.2% solution?

6. A treatment plant is adding 294 g/min of soda ash to its treated water. If the plant is producing water at 12 mgd, what is the soda ash usage in pounds per day and dosage in milligrams per liter?

7. What is the total head for a pump that is operating when the inlet pressure gauge reads 81 psi and the outlet gauge reads 124 psi?

8. Calculate the feed rate for sodium silicofluoride in gallons per day given the following data:

Flow rate is 38.5 mgd	Treated with 20.0% solution of H_2SiF_6
Fluoride desired is 1.00 mg/L	Fluoride ion percent is 79%
Fluoride in raw water is 0.30 mg/L	H_2SiF_6 weighs 9.8 lb/gal

9. What is the motor horsepower if 300 hp is required to run a pump with a motor efficiency of 90% and a pump efficiency of 83%?

10. A conventional treatment plant processes 2,450 gpm. If the lime dosage is 105 g/min, how many pounds of lime will the plant use in 1 month (exactly 30 days)? What is the average dosage in milligrams per liter?

Solutions to Water Treatment Grade 3 Test

1. First, find the length (in feet) of water filling the first casing that has a diameter of 1.5 ft.

Length of water filling the first casing = Depth of first casing – Depth of water to top of casing

Length of water filling the first casing = 100 ft – 74 ft = 26 ft

Then, determine the volume (in gallons) of water in the first casing using the following formula:

Volume, in gal = (0.785)(Diameter)^2(Length)(7.48 gal/ft^3)

Volume, in gal = (0.785)(1.5 ft)(1.5 ft)(26 ft)(7.48 gal/ft^3) = 344 gal

Next, find the number of gallons contained in the 1.0-ft diameter casing. Since the water level is above the 1.0-ft diameter casing, it follows that:

Depth of well, 284 ft – 100 ft for the top of 1.5 ft casing
= 184 ft of water-filled casing that is 1.0 ft in diameter

Using the above equation:

Volume, in gal = (0.785)(1.0 ft)(1.0 ft)(184 ft)(7.48 gal/ft^3) = 1,080 gal

Then, add the two volumes of water from each of the two casings.

344 gal + 1,080 gal = 1,424 gal

Next, determine the number of million gallons.

Million gallons = (1,424 gal)(1 M/1,000,000) = 0.001424 mil gal

(continued)

Last, using the "pounds" equation, calculate the number of pounds of calcium hypochlorite.

$$\text{Pounds, calcium hypochlorite} = \frac{(0.001424 \text{ mil gal})(50 \text{ mg/L})(8.34 \text{ lb/gal})}{(65\% \text{ Available chlorine}/100\%)}$$

Pounds, calcium hypochlorite = 0.914 lb, round to **0.9 lb of calcium hypochlorite**

2. First, find the capacity of the softener (in gallons).

The equation is:

$$\textbf{Water treatment capacity, gal} = \frac{\textbf{Exchange capacity, grains}}{\textbf{Hardness, gpg}}$$

$$\text{Water treatment capacity, gal} = \frac{7{,}498{,}000 \text{ grains}}{18.3 \text{ gpg}} = 409{,}727 \text{ gal}$$

Next, convert gallons per minute to gallons per hour.

$$(320 \text{ gpm})(60 \text{ min/hr}) = 19{,}200 \text{ gph}$$

Then, determine the operating time.

$$\textbf{Operating time, hr} = \frac{\textbf{Water treated, gal}}{\textbf{Flow rate, gph}}$$

$$\text{Operating time, hr} = \frac{409{,}727 \text{ gal}}{19{,}200 \text{ gph}} = 21.3 \text{ hr, round to } \mathbf{21 \text{ hr}}$$

3. First, find the length (in feet) of water in the casing.

Length of water-filled casing = Depth of well – Depth of water to top of casing

Length of water-filled casing = 306 ft – 114 ft = 192 ft

Then, convert the diameter from inches to feet.

$$\text{Diameter, ft} = \frac{14 \text{ in.}}{12 \text{ in./ft}} = 1.167 \text{ ft}$$

Next, determine the volume (in gallons) of water in the well casing using the following formula:

$$\textbf{Volume, in gal} = \mathbf{(0.785)(Diameter)^2(Length)(7.48\ gal/ft^3)}$$

$$\text{Volume, in gal} = (0.785)(1.167 \text{ ft})(1.167 \text{ ft})(192 \text{ ft})(7.48 \text{ gal/ft}^3) = 1{,}535 \text{ gal}$$

Next, determine the number of million gallons.

$$\text{Million gallons} = (1{,}535\text{ gal})(1\text{ M}/1{,}000{,}000) = 0.001535\text{ mil gal}$$

Last, using the "pounds" formula, calculate the number of pounds of calcium hypochlorite.

$$\text{Calcium hypochlorite, lb} = \frac{(0.001535\text{ mil gal})(50\text{ mg/L})(8.34\text{ lb/gal})}{(65\%\text{ Available chlorine}/100\%)}$$

Calcium hypochlorite, lb = 0.98 lb, round to **1 lb of calcium hypochlorite**

4. Flow in the 10-in. pipe equals flow in the 12-in. pipe because the flow must remain constant:

$$\mathbf{Q_1 = Q_2}$$

The equation is:

(Area 1)(Velocity 1) = (Area 2)(Velocity 2)

First, find the diameter for the 10-in. and 12-in. pipes (in feet).

$$\text{Diameter for 10-in.} = (10\text{-in.})(1\text{ ft}/12\text{ in.}) = 0.833\text{ ft}$$

$$\text{Diameter for 12-in.} = (12\text{-in.})(1\text{ ft}/12\text{ in.}) = 1.0\text{ ft}$$

Then, determine the area of each pipe.

$$\mathbf{Area = (0.785)(Diameter)^2}$$

$$\text{Area 1 (12-in.)} = (0.785)(1.0\text{ ft})(1.0\text{ ft}) = 0.785\text{ ft}^2$$

$$\text{Area 2 (10-in.)} = (0.785)(0.833\text{ ft})(0.833\text{ ft}) = 0.545\text{ ft}^2$$

Last, substitute areas calculated and known velocity in 10-in. pipe.

$$(0.785\text{ ft}^2)(x\text{, ft/s}) = (0.545\text{ ft}^2)(2.90\text{ ft/s})$$

Solve for x.

$$x\text{, ft/s} = \frac{(0.545\text{ ft}^2)(2.90\text{ ft/s})}{(0.785\text{ ft}^2)} = 2.01\text{ ft/s, round to } \mathbf{2\text{ ft/s in 12-in. pipe}}$$

5. Solve the problem using the dilution triangle.

12.8%	4.1	4.1 parts of the 12.8% solution are required for every 8.7 parts
8.2%		
4.1%	4.6	4.6 parts of the 4.1% solution are required for every 8.7 parts
	8.7 total parts	

$$\frac{4.1 \text{ parts (325 gal)}}{8.7 \text{ parts}} = \textbf{153 gal of the 12.8\% solution}$$

$$\frac{4.6 \text{ parts (325 gal)}}{8.7 \text{ parts}} = \frac{\textbf{172 gal of the 4.1\% solution}}{325 \text{ gal}}$$

To make 325 gal of the 8.2% solution, mix 153 gal of the 12.8% solution with 172 gal of the 4.1% solution.

6. First, find the pounds per day of soda ash usage.

Pounds per day, soda ash = (g/min)(1,440 min/d)(1 lb/454 g)

Pounds per day, soda ash = (294 g/min)(1,440 min/d)(1 lb/454 g)

= 933 lb/d of soda ash

Then, using the "pounds" equation, calculate the dosage in milligrams per liter.

$$\text{Dosage, mg/L} = \frac{933 \text{ lb/d}}{(12 \text{ mgd})(8.34 \text{ lb/gal})} = \textbf{9.3 mg/L of soda ash}$$

7. For a pump in operation:

Total head, ft = Static head, ft

The static head is the difference in the pressure gauges.

Static head, ft = (124 psi – 81 psi)(2.31 ft/psi)

= (43 psi)(2.31 ft/psi)

= 99.33 ft, round to **99 ft**

8. The equation is:

$$\textbf{F req. = F desired – F in raw water}$$

$$\text{F req.} = 1.00 \text{ mg/L} - 0.30 \text{ mg/L} = 0.70 \text{ mg/L}$$

$$\frac{(\%\text{ solution})(10{,}000\text{ mg/L})}{1\%} = \text{Number of mg/L}$$

$$20.0\%\text{ solution of } H_2SiF_6 = 200{,}000\text{ mg/L}$$

The equation is:

(F, mg/L)(mgd)(8.34 lb/gal) = (% as mg/L)(mgd flow)(Solution, lb/gal)(% F ion)

(0.7 mg/L)(38.5 mgd)(8.34 lb/gal) = (200,000 mg/L)(x mgd flow)(9.8 lb/gal)(79%/100% F)

Solve for x.

$$x\text{ mgd flow} = \frac{(0.70\text{ mg/L F})(38.5\text{ mgd})(8.34\text{ lb/gal})}{(200{,}000\text{ mg/L})(9.8\text{ lb/gal})(0.79\text{ F})} = 0.000145\text{ mgd}$$

Convert million gallons per day to gallons.

(0.000145 mgd)(1,000,000 gal/1 M) = 145 gal/d, round to **140 gal/d H_2SiF_6**

9. The equation is:

$$\textbf{Motor horsepower} = \frac{\textbf{(Water horsepower)}}{\textbf{(Motor efficiency)(Pump efficiency)}}$$

$$\text{Motor horsepower} = \frac{(300\text{ whp})}{(90\%/100\%\text{ Motor efficiency})(83\%/100\%\text{ Pump efficiency})}$$

$$= \frac{(300\text{ whp})}{(0.747)}$$

Motor horsepower = 401.6 mhp, round to **400 mhp**

10. First, convert gallons per minute to million gallons per day.

(2,450 gpm)(1,440 min/d)(1 M/1,000,000) = 3.528 mgd

Next, convert grams of lime per minute to pounds per day.

(105 g/min)(1 lb/454 g)(1,440 min/d) = 333 lb/d

(continued)

Then, determine the number of pounds used in 30 days.

$$(333 \text{ lb/d})(30 \text{ d/month}) = \mathbf{9{,}990 \text{ lb/month of lime}}$$

To find the dosage, use the "pounds" equation and rearrange to solve for dosage.

The equation is:

$$\textbf{Pounds per day} = \textbf{(mgd)(Dosage, mg/L)(8.34 lb/gal)}$$

Now, calculate the dosage.

$$\textbf{Dosage, mg/L} = \frac{\textbf{lb/d, lime}}{\textbf{(mgd)(8.34 lb/gal)}}$$

$$\text{Lime dosage, mg/L} = \frac{333 \text{ lb/d}}{(3.528 \text{ mgd})(8.34 \text{ lb/gal})}$$

$$= 11.32 \text{ mg/L, round to } \mathbf{11.3 \text{ mg/L of lim}}$$

CHAPTER

4

WATER TREATMENT

Grade 4

Students preparing for the grade 4 water treatment certification test should understand all of the problems presented for grades 1, 2, and 3.

FLOW AND VELOCITY PROBLEMS

1. Find the instantaneous flow in cubic feet per second for a trapezoidal channel that is 4 ft wide at the bottom and 6 ft wide at the water surface. The depth of the water is 44 in. and the flow velocity is 1.65 ft/s.

First, convert the depth of the water from inches to feet.

$$44 \text{ in.}/12 \text{ in. per ft} = 3.67 \text{ ft in depth}$$

The equation is:

$$\textbf{Flow, ft}^3\textbf{/s} = \frac{\textbf{(Bottom width + Water surface width)}}{\textbf{2}}\textbf{(Depth, ft)(Velocity, ft/s)}$$

$$\text{Flow, ft}^3\text{/s} = \frac{(6 \text{ ft} + 4 \text{ ft})(3.67 \text{ ft})(1.65 \text{ ft/s})}{2} = \mathbf{30\ ft^3/s}$$

2. **Water flowing through a full pipeline has a velocity of 3.2 ft/s. If the flow through the pipe is 1.74 ft^3/s, what is the diameter, in inches, of the pipeline?**

The equation is:

$$\textbf{Flow, ft}^3\textbf{/s} = \textbf{(Area, ft}^2\textbf{)(Velocity, ft/s)}$$

where the area = (0.785)(Diameter)^2

$$1.7 \text{ ft}^3\text{/s} = (0.785)(\text{Diameter})^2(3.2 \text{ ft/s})$$

Rearrange and solve for the diameter.

$$(\text{Diameter})^2 = \frac{1.74 \text{ ft}^3\text{/s}}{(0.785)(3.2 \text{ ft/s})} \quad (\text{Diameter})^2 = 0.6927 \text{ ft} \quad \text{Diameter} = 0.832 \text{ ft}$$

$$\text{D} = (0.832 \text{ ft})(12 \text{ in./ft}) = 9.984 \text{ in., round to } \mathbf{10 \text{ in.}}$$

3. **What is the velocity of flow, in feet per second, for a 4.0-in. diameter pipe if it delivers 148 gpm?**

First, convert the number of gallons per minute to cubic feet per second.

$$\text{Number of ft}^3\text{/s} = \frac{148 \text{ gpm}}{(7.48 \text{ gal/ft}^3)(60 \text{ s/min})} = 0.33 \text{ ft}^3\text{/s}$$

Next, convert the diameter from inches to feet.

$$\text{Number of ft} = (4.0 \text{ in.})(1 \text{ ft/12 in.}) = 0.333 \text{ ft}$$

The equation is:

$$\textbf{Flow, ft}^3\textbf{/s} = \textbf{(Area, ft}^2\textbf{)(Velocity, ft/s)}$$

where the area = (0.785)(Diameter)^2

$$0.33 \text{ ft}^3\text{/s} = (0.785)(0.333 \text{ ft})(0.333 \text{ ft})(\text{Flow, ft/s})$$

Rearrange and solve for the flow, in feet per second.

$$\text{Flow, ft/s} = \frac{0.33 \text{ ft}^3\text{/s}}{(0.785)(0.333 \text{ ft})(0.333 \text{ ft})} = 3.79 \text{ ft/s, round to } \mathbf{3.8 \text{ ft/s}}$$

4. Find the instantaneous flow, in cubic feet per second, for a trapezoidal channel that is 8.0 ft wide at the bottom and 12.0 ft wide at the water surface. The depth of the water is 67 in. and the flow velocity is 2.3 ft/s.

First, convert the depth in inches to feet.

$$67 \text{ in.}/12 \text{ in. per ft} = 5.58 \text{ ft}$$

The equation is:

$$\textbf{Flow, ft}^3\textbf{/s} = \frac{\textbf{(Bottom width)} + \textbf{(Water surface width)}}{\textbf{2}} \textbf{(Depth, ft)(Velocity, ft/s)}$$

$$\text{Flow, ft}^3\text{/s} = \frac{(8.0 \text{ ft} + 12.0 \text{ ft})}{2}(5.58 \text{ ft})(2.3 \text{ ft/s}) = 128.34 \text{ ft}^3\text{/s, round to } \mathbf{130 \text{ ft}^3\text{/s}}$$

5. Water flowing through a full pipeline has a velocity of 3.25 ft/s. If the flow through the pipe is 0.64 ft³/s, what is the diameter, in inches, of the pipeline?

The equation is:

$$\textbf{Flow, ft}^3\textbf{/s} = \textbf{(Area, ft}^2\textbf{)(Velocity, ft/s)}$$

where the area = (0.785)(Diameter)^2

$$0.64 \text{ ft}^3\text{/s} = (0.785)(\text{Diameter})^2(3.25 \text{ ft/s})$$

Rearrange and solve for the diameter.

$$(\text{Diameter})^2 = \frac{0.64 \text{ ft}^3\text{/s}}{(0.785)(3.25 \text{ ft/s})} \qquad (\text{Diameter})^2 = 0.25 \text{ ft} \qquad \text{Diameter} = 0.50 \text{ ft}$$

$$D = (0.50)(12 \text{ in./ft}) = \mathbf{6 \text{ in.}}$$

6. Water is flowing at a velocity of 2.4 ft/s in a 10.0-in. diameter pipe. If the pipe changes from 10.0 in. in diameter to 16 in., what will the velocity be in the 16-in. pipe?

Flow in the 10.0-in. pipe equals flow in the 16-in. pipe because flow must remain constant.

$$\mathbf{Q_1 = Q_2}$$

Because Q, Flow = (Area)(Velocity), it follows that:

$$\textbf{(Area 1)(Velocity 1)} = \textbf{(Area 2)(Velocity 2)}$$

First, find the diameter, in feet, of the 10.0-in. and 16-in. pipes.

$$\text{Diameter of 10.0-in. pipe} = 10.0 \text{ in.}(1 \text{ ft}/12 \text{ in.}) = 0.833 \text{ ft}$$

$$\text{Diameter of 16-in. pipe} = 16 \text{ in.}(1 \text{ ft}/12 \text{ in.}) = 1.33 \text{ ft}$$

Then, determine the area of each size pipe.

$$\text{Area} = (0.785)(\text{Diameter})^2$$

$$\text{Area 1 (10.0-in.)} = (0.785)(0.833 \text{ ft})(0.833 \text{ ft}) = 0.545 \text{ ft}^2$$

$$\text{Area 2 (16-in.)} = (0.785)(1.33 \text{ ft})(1.33 \text{ ft}) = 1.39 \text{ ft}^2$$

Last, substitute areas calculated and known velocity in 10.0-in. pipe.

$$(0.545 \text{ ft}^2)(2.4 \text{ ft/s}) = (1.39 \text{ ft}^2)(x, \text{ ft/s})$$

Solve for x.

$$x, \text{ ft/s} = \frac{(0.545 \text{ ft}^2)(2.4 \text{ ft/s})}{(1.39 \text{ ft}^2)} = \textbf{0.94 ft/s in 16-in. pipe}$$

7. Find the instantaneous flow, in cubic feet per second, for a trapezoidal channel that is 6 ft wide at the bottom and 9 ft wide at the water surface. The depth of the water is 65 in. and the flow velocity is 1.5 ft/s.

First, convert the depth of the water from inches to feet.

$$65 \text{ in.}/12 \text{ in. per ft} = 5.42 \text{ ft in depth}$$

The equation is:

$$\textbf{Flow, ft}^3\textbf{/s} = \frac{\textbf{(Width at bottom + Width at water surface)}}{\textbf{2}}\textbf{(Depth, ft)(Velocity, ft/s)}$$

$$\text{Flow, ft}^3\text{/s} = \frac{(6 \text{ ft} + 9 \text{ ft})}{2}(5.42 \text{ ft})(1.5 \text{ ft/s}) = \mathbf{61\ ft^3/s}$$

8. What is the velocity of flow, in feet per second, for a 14-in. diameter pipe if it delivers 1,040 gpm?

First, convert the number of gallons per minute to cubic feet per second.

$$\text{Number of ft}^3\text{/s} = \frac{1{,}040 \text{ gpm}}{(7.48 \text{ gal/ft}^3)(60 \text{ s/min})} = 2.32 \text{ ft}^3\text{/s}$$

Next, convert the diameter from inches to feet.

$$\text{Number of ft} = (14 \text{ in.})(1 \text{ ft}/12 \text{ in.}) = 1.167 \text{ ft}$$

The equation is:

$$\textbf{Flow, ft}^3\textbf{/s} = \textbf{(Area, ft}^2\textbf{)(Velocity, ft/s)}$$

where the area = $(0.785)(\text{Diameter})^2$

$$2.32 \text{ ft}^3\text{/s} = (0.785)(1.167 \text{ ft})(1.167 \text{ ft})(\text{Flow, ft/s})$$

Rearrange and solve for the flow, in feet per second.

$$\text{Flow, ft/s} = \frac{2.32 \text{ ft}^3\text{/s}}{(0.785)(1.167 \text{ ft})(1.167 \text{ ft})} = 2.17 \text{ ft/s, round to } \mathbf{2.2\ ft/s}$$

9. Water is flowing at a velocity of 3.0 ft/s in a 10.0-in. diameter pipe. If the pipe changes from 10.0 in. in diameter to 14 in., what will the velocity be in the 14-in. pipe?

Flow in the 10.0-in. pipe equals flow in the 14-in. pipe because flow must remain constant.

$$\mathbf{Q_1 = Q_2}$$

Because Q, Flow = (Area)(Velocity), it follows that:

(Area 1)(Velocity 1) = (Area 2)(Velocity 2)

First, find the diameter, in feet, of the 10.0-in. and 14-in. pipes.

$$\text{Diameter of 10.0-in. pipe} = 10.0 \text{ in.}(1 \text{ ft}/12 \text{ in.}) = 0.833 \text{ ft}$$

$$\text{Diameter of 14-in. pipe} = 14 \text{ in.}(1 \text{ ft}/12 \text{ in.}) = 1.167 \text{ ft}$$

Then, determine the area of each size pipe.

$$\text{Area} = (0.785)(\text{Diameter})^2$$

$$\text{Area 1 (10.0-in.)} = (0.785)(0.833 \text{ ft})(0.833 \text{ ft}) = 0.545 \text{ ft}^2$$

$$\text{Area 2 (14-in.)} = (0.785)(1.167 \text{ ft})(1.167 \text{ ft}) = 1.069 \text{ ft}^2$$

Last, substitute areas calculated and known velocity in 10-in. pipe.

$$(0.545 \text{ ft}^2)(3.0 \text{ ft/s}) = (1.069 \text{ ft}^2)(x, \text{ ft/s})$$

Solve for x.

$$x, \text{ ft/s} = \frac{(0.545 \text{ ft}^2)(3.0 \text{ ft/s})}{(1.069 \text{ ft}^2)} = \mathbf{1.5 \text{ ft/s in 14-in. pipe}}$$

10. Water is flowing at a velocity of 2.75 ft/s in a 12-in. diameter pipe. If the pipe changes from 12 in. in diameter to 16 in., what will the velocity be in the 16-in. pipe?

Flow in the 10-in. pipe equals flow in the 14-in. pipe because flow must remain constant.

$$\mathbf{Q_1 = Q_2}$$

Because Q, Flow = (Area)(Velocity), it follows that:

(Area 1)(Velocity 1) = (Area 2)(Velocity 2)

First, find the diameter, in feet, of the 10-in. and 14-in. pipes.

Diameter of 12-in. pipe, in ft = 12 in.(1 ft/12 in.) = 1.0 ft

Diameter of 16-in. pipe, in ft = 16 in.(1 ft/12 in.) = 1.33 ft

Then, determine the areas of each size pipe.

$$\text{Area} = (0.785)(\text{Diameter})^2 \text{ (or use } \pi r^2)$$

$$\text{Area 1 (12-in.)} = (0.785)(1.0 \text{ ft})(1.0 \text{ ft}) = 0.785 \text{ ft}^2$$

$$\text{Area 2 (16-in.)} = (0.785)(1.33 \text{ ft})(1.33 \text{ ft}) = 1.39 \text{ ft}^2$$

Last, substitute areas calculated and known velocity in 12-in. pipe.

$$(0.785 \text{ ft}^2)(2.75 \text{ ft/s}) = (1.39 \text{ ft}^2)(x, \text{ ft/s})$$

Solve for x.

$$x, \text{ ft/s} = \frac{(0.785 \text{ ft}^2)(2.75 \text{ ft/s})}{(1.39 \text{ ft}^2)} = 1.55 \text{ ft/s, round to } \mathbf{1.6 \text{ ft/s in 16-in. pipe}}$$

PRESSURE PROBLEMS

11. A tank is 60.0 ft in diameter and 24.0 ft tall. If there are 240,500 gal of water in the tank, what is the pounds per square inch at the bottom of the tank? What is the pounds per square inch 5.0 ft above the bottom of the tank?

First, calculate the amount of water present in cubic feet.

$$\frac{240{,}500 \text{ gal}}{7.48 \text{ gal/ft}^3} = 32{,}152 \text{ ft}^3 \text{ then, the number of ft}^3 = \pi r^2(\text{Depth})$$

$$\text{Radius} = \text{Diameter}/2 = 60.0/2 = 30.0 \text{ ft}$$

$$32{,}152 \text{ ft}^3 = 3.14(30.0 \text{ ft})(30.0 \text{ ft})(\text{Depth})$$

$$\text{Depth} = \frac{32{,}152 \text{ ft}^3}{3.14(30.0 \text{ ft})(30.0 \text{ ft})}$$

$$\text{Depth} = 11.38 \text{ ft}$$

Now, solve for the number of pounds per square inch at the bottom of the tank.

$$\text{psi} = \frac{\text{Depth}}{2.31 \text{ ft/psi}} = \frac{11.38 \text{ ft}}{2.31 \text{ ft/psi}} = \mathbf{4.93 \text{ psi}}$$

Solve for pounds per square inch 5 ft above the bottom of the tank. The depth of the water is 5 ft less than the total depth (11.38 ft – 5.0 ft = 6.38 ft). Thus:

$$\text{psi} = \frac{\text{Depth}}{2.31 \text{ ft/psi}} = \frac{6.38 \text{ ft}}{2.31 \text{ ft/psi}} = 2.76 \text{ psi, round to } \mathbf{2.8 \text{ psi}}$$

12. A tank is 30.0 ft in diameter and 20 ft tall. If there are 78,500 gal of water in the tank, what is the pounds per square inch at the bottom of the tank? What is the pounds per square inch 5.0 ft above the bottom of the tank?

First, calculate the number of cubic feet of water present.

$$\frac{78{,}500 \text{ gal}}{7.48 \text{ gal/ft}^3} = 10{,}495 \text{ ft}^3$$

Then, determine the radius.

$$r = \text{Diameter}/2 = 30.0/2 = 15.0 \text{ ft}$$

Solve for depth.

$$10{,}495 \text{ ft}^3 = 3.14(15.0 \text{ ft})(15.0 \text{ ft})(\text{Depth})$$

$$\text{Depth} = \frac{10{,}495 \text{ ft}^3}{3.14(15.0 \text{ ft})(15.0 \text{ ft})}$$

$$\text{Depth} = 14.85 \text{ ft}$$

Now, solve for the number of pounds per square inch at the bottom of the tank.

$$\text{psi} = \frac{\text{Depth}}{2.31 \text{ ft/psi}} = \frac{14.85 \text{ ft}}{2.31 \text{ ft/psi}} = \mathbf{6.43 \text{ psi}}$$

Then, solve for pounds per square inch 5 ft above the bottom of the tank. The depth of the water is 5 ft less than the total depth (14.85 ft – 5.0 ft = 9.85 ft). Thus:

$$\text{psi} = \frac{\text{Depth}}{2.31 \text{ ft/psi}} = \frac{9.85 \text{ ft}}{2.31 \text{ ft/psi}} = 4.26 \text{ psi, round to } \mathbf{4.3 \text{ psi}}$$

DETENTION PROBLEMS

13. Calculate the theoretical detention time, in hours, for the following water treatment plant:

Flow rate of 18.1 mgd.

Five flocculation basins measuring 48.0 ft by 10.0 ft by 10.0 ft in average depth each.

Sedimentation (sed) basin measuring 398 ft by 62.0 ft by 10.5 ft in average depth.

Eight filters measuring 40.0 ft by 28.0 ft by 12.0 ft in depth each.

Clear well averages 1.85 mil gal.

First, determine the capacity, in gallons, for each basin by converting from million gallons to gallons. Then, add amounts for each basin for total volume, in gallons.

Volume, gal, of floc basins = (48.0 ft)(10.0 ft)(10.0 ft)(7.48 gal/ft^3)(5 basins) = 179,520 gal

Volume of sed basin = (398 ft)(62.0 ft)(10.5 ft)(7.48 gal/ft^3) = 1,938,053 gal

Volume of filters = (40.0 ft)(28.0 ft)(12.0 ft)(7.48 gal/ft^3)(8 filters) = 804,250 gal

Volume of clear well = (1.85 mil gal)(1,000,000/1 M) = 1,850,000 gal

4,771,823 gal

Then, convert the flow rate of 18.1 mgd to gallons per hour.

$$\text{Gallons per hour} = (18.1 \text{ mgd})\frac{(1{,}000{,}000 \text{ gal})}{1 \text{ M}}\frac{(1 \text{ d})}{24 \text{ hr}} = 754{,}167 \text{ gph}$$

Write the equation with units asked for in question:

$$\textbf{Detention time, hr} = \frac{\textbf{Volume, gal}}{\textbf{Flow rate, gph}}$$

Substituting in equation above:

$$\text{Detention time, hr} = \frac{4{,}771{,}823 \text{ gal}}{754{,}167 \text{ gph}} = \mathbf{6.33 \text{ hr}}$$

14. Calculate the theoretical detention time, in hours, for the following treatment plant:

Flow rate of 45 mgd.

Clear well with a 15-mil gal capacity.

The following measurements were made to the nearest foot (two significant figures at least):

Four flocculation basins measuring 60 ft by 10 ft by 10 ft each.

A sedimentation (sed) basin measuring 350 ft by 80 ft by 12 ft.

Twelve filters measuring 40 ft by 30 ft by 15 ft each.

First, determine the capacity, in gallons, for each basin by converting from million gallons to gallons. Then, add the amounts for each basin for total volume, in gallons.

Volume, gal, of floc basins = (60 ft)(10 ft)(10 ft)(7.48 gal/ft^3)(4 basins) = 179,520 gal

Volume of sed basin = (350 ft)(80 ft)(12 ft)(7.48 gal/ft^3) = 2,513,280 gal

Volume of filters = (40 ft)(30 ft)(15 ft)(7.48 gal/ft^3)(12 filters) = 1,615,680 gal

Volume of clear well = (15 mil gal)(1,000,000/1 M) = 15,000,000 gal

19,308,480 gal

Then, convert the flow rate of 45 mgd to gallons per hour.

$$\text{Gallons per hour} = (45\ \text{mgd})\frac{(1{,}000{,}000\ \text{gal})\ (1\ \text{d})}{1\ \text{M} \qquad 24\ \text{hr}} = 1{,}875{,}000\ \text{gph}$$

Substitute known values:

$$\textbf{Detention time, hr} = \frac{\textbf{Volume, gal}}{\textbf{Flow rate, gph}}$$

$$\text{Detention time, hr} = \frac{19{,}308{,}480\ \text{gal}}{1{,}875{,}000\ \text{gph}} = 10.29\ \text{hr, round to } \mathbf{10\ hr}$$

15. Calculate the theoretical detention time, in hours, for the following treatment plant:

Flow rate of 33 mgd.

Six flocculation basins measuring 40.0 ft by 10.0 ft by 10.0 ft each.

A sedimentation (sed) basin measuring 400.0 ft by 50.0 ft by 10.0 ft.

Twelve filters measuring 40.0 ft by 30.0 ft by 12.0 ft each.

Clear well with 7.9 mil gal.

First, determine the capacity, in gallons, for each basin by converting from million gallons to gallons. Then, add the amounts for each basin for total volume, in gallons.

Volume, gal, of floc basins = (40.0 ft)(10.0 ft)(10.0 ft)(7.48 gal/ft^3)(6 basins) = 179,520 gal

Volume of sed basin = (400.0 ft)(50.0 ft)(10.0 ft)(7.48 gal/ft^3) = 1,496,000 gal

Volume, filters = (40.0 ft)(30.0 ft)(12.0 ft)(7.48 gal/ft^3)(12 filters) = 1,292,544 gal

Volume of clear well = (7.9 mil gal)(1,000,000/1 M) = 7,900,000 gal

Total: 10,868,064 gal

Then, convert the flow rate of 33 mgd to gallons per hour.

$$\text{Gallons per hour} = (33\text{ mgd})\frac{(1{,}000{,}000\text{ gal})\ (1\text{ d})}{1\text{ M}\qquad 24\text{ hr}} = 1{,}375{,}000\text{ gph}$$

Substitute known values:

$$\textbf{Detention time, hr} = \frac{\textbf{Volume, gal}}{\textbf{Flow rate, gph}}$$

$$\text{Detention time, hr} = \frac{10{,}868{,}064\text{ gal}}{1{,}375{,}000\text{ gph}} = \mathbf{7.9\ hr}$$

16. A backwash tank has a volume of 550,000 gal. It currently has 270,000 gal left over from the previous day's backwashes because the power to the three recirculation pumps failed. The power has now been restored, but one of the pumps does not work. The operator has determined that six filters require backwashing for the shift. Each filter takes 80,000 gal to backwash. If the two remaining recirculation pumps are removing the backwash water at a rate of 20,000 gal/hr and each backwash takes 1 hr and 10 min, will the operator have enough room such that no waiting will be required?

First, find the number of hours it will take to do six backwashes.

$$\text{(Backwash 6 filters)(1 hr 10 min/filter)} = \text{7 hr}$$

Next, determine the number of gallons the recirculation pumps will remove during 7 hr of backwashing.

$$\text{(20,000 gal/hr)(7 hr)} = \text{140,000 gal}$$

If all six filters were backwashed without waiting, would there be enough room in the recirculation tank? To determine this, add the previous day's water left in the tank and subtract what the recirculation pumps would remove.

$$\text{(6 filters)(80,000 gal/filter)} + \text{270,000 gal} - \text{140,000 gal} = \text{610,000 gal}$$

$$\text{610,000 gal} - \text{550,000 gal capacity of tank} = \textbf{60,000 gal over}$$

Clearly, the operator is going to either wait before doing the last backwash or only backwash five filters that day.

DOSAGE PROBLEMS

17. How many pounds per day of calcium hypochlorite are needed to treat 7.1 mgd with a dosage of 2.0 mg/L if the calcium hypochlorite has 62% available chlorine?

Using the "pounds" equation:

$$\textbf{Number of lb/d} = \textbf{(mgd)(Dosage, mg/L)(8.34 lb/gal)}$$

$$\text{Pounds per day} = (7.1\text{ mgd})(2.0\text{ mg/L})(8.34\text{ lb/gal}) = 118.428\text{ lb/d of chlorine}$$

$$\frac{118.428\text{ lb Chlorine required}}{62\%\text{ Available chlorine}/100\%} = 191.01\text{ lb, round to } \textbf{190 lb of calcium hypochlorite}$$

18. A water treatment plant has a flow of 25 mgd and is being treated with 4,010 gpd of a hypochlorite solution. If the desired dose is 2.5 mg/L, determine the concentration of the hypochlorite solution in percent.

First, convert 4,010 gpd of the hypochlorite solution rate of flow to million gallons per day.

$$(4{,}010\text{ gpd})\frac{(1\text{ mil gal})}{(1{,}000{,}000\text{ gal})} = 0.00401\text{ mgd}$$

Then, using the equation:

$$\mathbf{C_1V_1 = C_2V_2}$$

$$(x\text{ mg/L})(0.00401\text{ mgd}) = (25\text{ mgd})(2.5\text{ mg/L})$$

$$x\text{ mg/L, hypochlorite} = \frac{(25\text{ mgd})(2.5\text{ mg/L})}{(0.00401\text{ mgd})} = 15{,}586\text{ mg/L}$$

The equation is:

$$1\% = 10{,}000\text{ mg/L}$$

Last, convert mg/L into percent.

$$\frac{(15{,}586\text{ mg/L})\quad(1\%)}{10{,}000\text{ mg/L}} = 1.5586\%\text{, round to } \textbf{1.6\% hypochlorite solution}$$

19. A water treatment plant has a filter effluent flow of 5,400 gpm and is being treated with 850 gpd of a hypochlorite solution. If the desired dose is 2.25 mg/L, determine the concentration of the hypochlorite solution in percent.

First, convert 5,400 gpm filter flow to million gallons per day.

$$\frac{(5{,}400 \text{ gpm})\ (1{,}440 \text{ min})\ (1 \text{ mil gal})}{\text{day} \quad 1{,}000{,}000 \text{ gal}} = 7.776 \text{ mgd}$$

Then, convert 850 gpd of the hypochlorite solution rate of flow to million gallons per day.

$$\frac{(850 \text{ gpd})\ (1 \text{ mil gal})}{1{,}000{,}000 \text{ gal}} = 0.00085 \text{ mgd}$$

Then, using the equal dosage equation:

$$(0.00085 \text{ mgd})(x \text{ mg/L})(8.34 \text{ lb/gal}) = (7.776 \text{ mgd})(2.25 \text{ mg/L})(8.34 \text{ lb/gal})$$

$$x \text{ mg/L} = \frac{(7.776 \text{ mgd})(2.25 \text{ mg/L})(8.34 \text{ lb/gal})}{(0.00085 \text{ mgd})(8.34 \text{ lb/gal})} = 20{,}584 \text{ mg/L}$$

Last, convert milligrams per liter to percent.

$$\frac{(20{,}584 \text{ mg/L}) \quad (1\%)}{10{,}000 \text{ mg/L}} = 2.0584\%$$, round to **2.1% hypochlorite solution**

20. A water treatment plant has a filter flow of 3,250 gpm and is being treated with 650 gpd of a hypochlorite solution. If the desired dose is 3.0 mg/L, determine the concentration of the hypochlorite solution in percent.

First, convert 3,250 gpm filter flow to million gallons per day.

$$\frac{(3{,}250 \text{ gpm})\ (1{,}440 \text{ min})\ (1 \text{ mil gal})}{\text{day}\quad 1{,}000{,}000 \text{ gal}} = 4.68 \text{ mgd}$$

Then, convert 650 gpd of the hypochlorite solution rate of flow to million gallons per day.

$$\frac{(650 \text{ gpd})\ (1 \text{ mil gal})}{1{,}000{,}000 \text{ gal}} = 0.00065 \text{ mgd}$$

Next, using the equal dosage equations:

$$(0.00065 \text{ mgd})(x \text{ mg/L})(8.34 \text{ lb/gal}) = (4.68 \text{ mgd})(3.0 \text{ mg/L})(8.34 \text{ lb/gal})$$

$$\text{mg/L, hypochlorite} = \frac{(4.68 \text{ mgd})(3.0 \text{ mg/L})(8.34 \text{ lb/gal})}{(0.00065 \text{ mgd})(8.34 \text{ lb/gal})} = 21{,}600 \text{ mg/L}$$

Last, convert milligrams per liter to percent.

$$\frac{(21{,}600 \text{ mg/L})\quad (1\%)}{10{,}000 \text{ mg/L}} = 2.16\%, \text{ round to } \mathbf{2.2\%\ hypochlorite\ solution}$$

21. A water treatment plant is treating 14.5 mgd with 274 lb/d of chlorine. What is the chlorine dosage in milligrams per liter?

The equation is:

Pounds per day, chlorine = (mgd)(Dosage, mg/L)(8.34 lb/gal)

274 lb/d = (14.5 mgd)(Dosage, mg/L)(8.34 lb/gal)

Solving for dosage:

$$\text{Dosage, mg/L} = \frac{274 \text{ lb/d}}{(14.5 \text{ mgd})(8.34 \text{ lb/gal})} = 2.266 \text{ mg/L, round to } \mathbf{2.26\ mg/L,\ chlorine}$$

22. Determine the pounds per day of chlorine required if a treatment plant is treating 12.25 mgd with a dosage of 2.20 mg/L.

The equation is:

$$\textbf{Pounds per day, chlorine} = \textbf{(mgd)(Dosage, mg/L)(8.34 lb/gal)}$$

$$\text{Pounds per day, chlorine} = (12.25 \text{ mgd})(2.20 \text{ mg/L})(8.34 \text{ lb/gal})$$

$$x \text{ pounds per day, chlorine} = 224.76 \text{ lb/d, round to } \textbf{225 lb/d of chlorine}$$

23. A 5-min drawdown test result shows that 108 mL of a cationic polymer are being used to treat the raw water. The specific gravity (sp gr) of the polymer is 1.18. If the plant is treating 5,370 gpm, what is the polymer dosage in milligrams per liter?

First, determine the number of milliliters per minute for the polymer.

$$\frac{108 \text{ mL}}{5 \text{ min}} = 21.6 \text{ mL/min}$$

Next, calculate the pounds per gallon for the polymer.

$$\textbf{Pounds per gallon} = \textbf{(Specific gravity)(8.34 lb/gal)}$$

$$\text{lb/gal, polymer} = (1.18)(8.34 \text{ lb/gal}) = 9.84 \text{ lb/gal}$$

Then, find the number of million gallons per day.

$$(5{,}370 \text{ gpm})(1{,}440 \text{ min/d})(1 \text{ M}/1{,}000{,}000) = 7.73 \text{ mgd}$$

Use the dosage equation with conversions added for convenience (dosage/conversion equation):

$$\textbf{mg/L, polymer} = \frac{\textbf{(mL/min)(1,440 min/d)(lb/gal, polymer)}}{\textbf{(3,785 mL/gal)(mgd)(8.34 lb/gal)}}$$

$$\text{mg/L, polymer} = \frac{(21.6 \text{ mL/min})(1{,}440 \text{ min/d})(9.84 \text{ lb/gal})}{(3{,}785 \text{ mL/gal})(7.73 \text{ mgd})(8.34 \text{ lb/gal})} = \textbf{1.25 mg/L of polymer}$$

24. A plant treats 16.2 mgd with a cationic polymer that weighs 10.27 lb/gal. If a drawdown test gives a result of 280 mL in 5 min, what is the polymer dosage in milligrams per liter?

First, determine the number of milliliters per minute for the polymer.

$$\frac{280 \text{ mL}}{5 \text{ min}} = 56 \text{ mL/min}$$

Dosage/conversion equation:

$$\textbf{mg/L, polymer} = \frac{\textbf{(mL/min)(1,440 min/d)(lb/gal, polymer)}}{\textbf{(3,785 mL/gal)(mgd)(8.34 lb/gal)}}$$

$$\text{mg/L, polymer} = \frac{(56 \text{ mL/min})(1{,}440 \text{ min/d})(10.27 \text{ lb/gal})}{(3{,}785 \text{ mL/gal})(16.2 \text{ mgd})(8.34 \text{ lb/gal})} = \textbf{1.6 mg/L of polymer}$$

25. How many pounds per day of 60.0% calcium hypochlorite are required to maintain a 2.5-mg/L dosage for a 7,500-gpm treatment plant?

First, determine the number of million gallons per day.

$$\text{Million gallons per day} = \frac{(7{,}500 \text{ gpm})(1{,}440 \text{ min/d}) \quad (1 \text{ M})}{1{,}000{,}000} = 10.8 \text{ mgd}$$

Pounds per day, chlorine = (mgd)(Dosage, mg/L)(8.34 lb/gal)

Pounds per day, chlorine = (10.8 mgd)(2.5 mg/L)(8.34 lb/gal) = 225.18 lb/d required

The calcium hypochlorite is not 100% pure, which means you need to divide by the percent purity, 60.0%.

$$\frac{225.18 \text{ lb/d required}}{60.0\% \text{ Available } Cl_2 / 100\%} = 375.3 \text{ lb/d, round to } \textbf{380 lb/d of calcium hypochlorite}$$

26. Calculate the potassium permanganate ($KMnO_4$) dosage if, after aeration, 0.25 mg/L iron (Fe) and 0.28 mg/L of manganese (Mn) remain.

The equation is:

$$\mathbf{KMnO_4,\ mg/L = 0.2(Fe,\ mg/L) + 2.0(Mn,\ mg/L)}$$

$$KMnO_4,\ mg/L = 0.2(0.25,\ mg/L\ of\ Fe) + 2.0(0.28,\ mg/L\ of\ Mn)$$

$$KMnO_4,\ mg/L = 0.05\ mg/L + 0.56\ mg/L$$

$$KMnO_4,\ mg/L = \mathbf{0.61\ mg/L}$$

27. A 5-min drawdown test result shows that 110 mL of 48% alum are being injected at the rate of 10.43 lb/gal. If the plant is treating 875 gpm, what is the alum dosage in milligrams per liter?

First, determine the number of milliliters per minute for the alum.

$$\frac{110\ mL}{5\ min} = 22\ mL/min$$

Next, find the number of milligrams per liter being treated.

$$(875\ gpm)(1{,}440\ min/d)(1\ M/1{,}000{,}000) = 1.26\ mgd$$

The dosage/conversion equation is:

$$\mathbf{mg/L,\ alum = \frac{(mL/min)(1{,}440\ min/d)(lb/gal,\ alum)}{(3{,}785\ mL/gal)(mgd)(8.34\ lb/gal)(Purity\ of\ alum)}}$$

$$mg/L,\ alum = \frac{(22\ mL/min)(1{,}440\ min/d)(10.43\ lb/gal)}{(3{,}785\ mL/gal)(1.26\ mgd)(8.34\ lb/gal)(48\%/100\%)}$$

$$= \mathbf{17\ mg/L\ of\ alum}$$

28. A 10-min drawdown test result shows that 535 mL of alum that is 45.0% pure are being injected at a rate of 9.98 lb/gal. If the plant is treating 1,250 gpm, what is the alum dosage in milligrams per liter?

First, determine the number of milliliters per minute for the alum.

$$\frac{535 \text{ mL}}{10 \text{ min}} = 53.5 \text{ mL/min}$$

Next, find the number of million gallons per day being treated.

$$(1{,}250 \text{ gpm})(1{,}440 \text{ min/d})(1 \text{ M}/1{,}000{,}000) = 1.80 \text{ mgd}$$

The dosage/conversion equation is:

$$\textbf{mg/L, alum} = \frac{\textbf{(mL/min)(1,440 min/d)(lb/gal, alum)}}{\textbf{(3,785 mL/gal)(mgd)(8.34 lb/gal)(Purity of alum)}}$$

$$\text{mg/L, alum} = \frac{(53.5 \text{ mL/min})(1{,}440 \text{ min/d})(9.98 \text{ lb/gal})}{(3{,}785 \text{ mL/gal})(1.80 \text{ mgd})(8.34 \text{ lb/gal})(45.0\%/100\%)}$$

= **30.1 mg/L of alum**

29. A plant uses a coagulant aid that weighs 10.21 lb/gal to treat 4.4 mgd. The results of a drawdown test show that 127 mL of coagulant aid are used in 5 min. What is the coagulant aid dosage in milligrams per liter?

First, determine the number of milliliters per minute for the coagulant aid.

$$\frac{127 \text{ mL}}{5 \text{ min}} = 25.4 \text{ mL/min}$$

The dosage equation is:

$$\textbf{Coagulant aid, mg/L} = \frac{\textbf{(mL/min)(1,440 min/d)(lb/gal, Coagulant aid)}}{\textbf{(3,785 mL/gal)(mgd)(8.34 lb/gal)}}$$

$$\text{Coagulant aid, mg/L} = \frac{(25.4 \text{ mL/min})(1{,}440 \text{ min/d})(10.21 \text{ lb/gal})}{(3{,}785 \text{ mL/gal})(4.4 \text{ mgd})(8.34 \text{ lb/gal})}$$

= **2.7 mg/L of coagulant aid**

30. How many fluid ounces of sodium hypochlorite (5.5% available chlorine) are required to disinfect a well with the following parameters: (1) depth of well is 185.5 ft; (2) 12-in. diameter well casing extends down to 100.0 ft; (3) the remainder is an 8.0-in. diameter casing; the desired dose is 50.0 mg/L; and (4) the depth to water is 50.1 ft. Assume the sodium hypochlorite solution weighs 8.99 lb/gal.

First, find the diameter, in feet, for both well casings.

$$\text{Diameter for 12-in. casing} = \frac{(12\text{ in.})(1\text{ ft})}{12\text{ in.}} = 1.0\text{ ft}$$

$$\text{Diameter for 8-in. casing} = \frac{(8\text{ in.})(1\text{ ft})}{12\text{ in.}} = 0.667\text{ ft}$$

Then, find the length, in feet, of the water column in the casing.

Length of water-filled casing = Depth of well – Depth of water to top of casing

Length of water-filled casing for 12-in. diameter = 100.0 ft – 50.1 ft = 49.9 ft

Length of water-filled casing for 8-in. diameter = 185.5 ft – 100.0 ft = 85.5 ft

Next, determine the volume, in gallons, in the well casings using the following formula:

Volume, in gal = (0.785)(Diameter)^2(Length)(7.48 gal/ft^3)

Volume, in gal, for 12-in. casing = (0.785)(1.0 ft)(1.0 ft)(49.9 ft)(7.48 gal/ft^3)
= 293.00 gal

Volume, in gal, for 8-in. casing = (0.785)(0.667 ft)(0.667 ft)(85.5 ft)(7.48 gal/ft^3)
= 223.35 gal

Total water in well casings = 293.00 gal + 223.35 gal = 516.35 gal

Next, determine the number of million gallons.

Million gallons = (516.35 gal)(1 M/1,000,000) = 0.00051635 mil gal

Then, using the "pounds" equation, calculate the number of pounds of sodium hypochlorite.

$$\text{Pounds, sodium hypochlorite} = \frac{(0.00051635\text{ mil gal})(50.0\text{ mg/L})(8.34\text{ lb/gal})}{5.5\%\text{ Available chlorine}/100\%}$$

Pounds, sodium hypochlorite = 3.91 lb

Next, calculate the number of gallons of sodium hypochlorite required.

(3.91 lb)/(8.99 lb/gal) = 0.435 gal

Last, convert to fluid ounces.

0.435 gal(128 oz/gal) = 55.7 oz, round to **56 oz of sodium hypochlorite**

31. How many fluid ounces of sodium hypochlorite (5.3% available chlorine) are required to disinfect a well with the following parameters: (1) depth of well is 276 ft; (2) 14-in. diameter well casing extends down 100.0 ft; (3) remainder of casing is 10 in. in diameter; (4) the residual desired dose is 50.0 mg/L; (5) the depth to water is 64.3 ft; and (6) the chlorine demand is 17 mg/L? Assume the sodium hypochlorite solution weighs 8.95 lb/gal.

First, find the diameter, in feet, for both well casings.

$$\text{Diameter for 14-in. casing} = \frac{(14\text{ in.})(1\text{ ft})}{12\text{ in.}} = 1.167\text{ ft}$$

$$\text{Diameter for 10-in. casing} = \frac{(10\text{ in.})(1\text{ ft})}{12\text{ in.}} = 0.833\text{ ft}$$

Then, find the length, in feet, of the water column in the casing.

Length of water-filled casing = Depth of given diameter casing – Depth of water to top of casing

Length of water-filled casing for 14-in. diameter = 100.0 ft – 64.3 ft = 35.7 ft

Length of water-filled casing for 10-in. diameter = 276 ft – 100 ft = 176 ft

Next, determine the volume, in gallons, in the well casings using the following formula:

Volume, in gal = $(0.785)(\text{Diameter})^2(\text{Length})(7.48\text{ gal/ft}^3)$

Volume, in gal, 14-in. = $(0.785)(1.167\text{ ft})(1.167\text{ ft})(35.7\text{ ft})(7.48\text{ gal/ft}^3)$

= 285.48 gal

Volume, in gal, 10-in. = $(0.785)(0.833\text{ ft})(0.833\text{ ft})(176\text{ ft})(7.48\text{ gal/ft}^3)$

= 717.09 gal

Total water in well casings = 285.48 gal + 717.09 gal = 1,002.57 gal

Next, determine the number of million gallons.

Million gallons = (1,002.57 gal)(1 M/1,000,000) = 0.001 mil gal

Find the chlorine required.

Chlorine required = Chlorine demand + Chlorine residual

Cl_2 req. = 17 mg/L + 50 mg/L = 67 mg/L

Then, using the "pounds" equation, calculate the number of pounds of sodium hypochlorite.

$$\text{Pounds, sodium hypochlorite} = \frac{(0.001\text{ mil gal})(67\text{ mg/L})(8.34\text{ lb/gal})}{5.3\%\text{ Available chlorine}/100\%}$$

$$\text{Pounds, sodium hypochlorite} = 10.54\text{ lb}$$

Next, calculate the number of gallons of sodium hypochlorite required.

$$(10.54\text{ lb})/(8.95\text{ lb/gal}) = 1.178\text{ gal}$$

Last, convert to fluid ounces.

1.178 gal(128 oz/gal) = 150.78 oz, round to **150 oz of sodium hypochlorite**

32. A dosage of 0.30 mg/L of copper sulfate pentahydrate is needed to control algae in a 350-mil gal capacity lake. If the available copper (Cu) is 25%, how many pounds of copper sulfate pentahydrate are required?

$$\textbf{Copper sulfate, lb} = \frac{\textbf{(mil gal)(Dosage, mg/L)(8.34 lb/gal)}}{\textbf{\% Available Cu/100\%}}$$

$$\text{Copper sulfate, lb} = \frac{(350\text{ mgd})(0.30\text{ mg/L})(8.34\text{ lb/gal})}{25\%/100\%}$$

= 3,503 lb, round to **3,500 lb of copper sulfate**

33. A 5-min drawdown test shows that 67 mL of a cationic polymer are being injected. The specific gravity of the polymer is 1.22. If the plant is treating 4,875 gpm, what is the polymer dosage in milligrams per liter?

First, determine the number of milliliters per minute for the polymer.

$$\frac{67 \text{ mL}}{5 \text{ min}} = 13.4 \text{ mL/min}$$

Next, calculate the pounds per gallon for the polymer.

$$\textbf{Pounds per gallon} = \textbf{(Specific gravity)(8.34 lb/gal)}$$

$$\text{lb/gal, polymer} = (1.22 \text{ sp gr})(8.34 \text{ lb/gal}) = 10.175 \text{ lb/gal}$$

Then, find the number of million gallons per day.

$$(4{,}875 \text{ gpm})(1{,}440 \text{ min/d})(1 \text{ M}/1{,}000{,}000) = 7.02 \text{ mgd}$$

The dosage/conversion equation is:

$$\textbf{mg/L, Polymer} = \frac{\textbf{(mL/min)(1,440 min/d)(lb/gal, polymer)}}{\textbf{(3,785 mL/gal)(mgd)(8.34 lb/gal)}}$$

$$\text{mg/L, polymer} = \frac{(13.4 \text{ mL/min})(1{,}440 \text{ min/d})(10.175 \text{ lb/gal})}{(3{,}785 \text{ mL/gal})(7.02 \text{ mgd})(8.34 \text{ lb/gal})} = \textbf{0.89 mg/L of polymer}$$

34. A dosage of 0.50 mg/L of copper sulfate pentahydrate is desired to control algae in a 122 acre-ft capacity reservoir. If the available copper is 25%, how many pounds of copper sulfate pentahydrate are required?

First, convert the volume of the reservoir to million gallons.

$$(43{,}560 \text{ ft}^3\text{/acre-ft})(122 \text{ acre-ft})(7.48 \text{ gal/ft}^3)(1 \text{ M}/1{,}000{,}000) = 39.75 \text{ mil gal}$$

The equation is:

$$\textbf{Copper sulfate, lb} = \frac{\textbf{(mil gal)(Dosage, mg/L)(8.34 lb/gal)}}{\textbf{\% Available Cu/100\%}}$$

$$\text{Copper sulfate, lb} = \frac{(39.75 \text{ mgd})(0.50 \text{ mg/L})(8.34 \text{ lb/gal})}{25\%/100\%}$$

$$= 663 \text{ lb, round to } \textbf{660 lb of copper sulfate}$$

35. A dosage of 0.10 mg/L of copper sulfate pentahydrate is needed in a reservoir that contains trout (trout are sensitive to copper sulfate, and it has been shown that dosages over 0.14 mg/L are toxic to this fish). The capacity of the reservoir is 22,000 acre-ft. If the available copper is 25%, how many pounds of copper sulfate pentahydrate are needed?

First, convert the volume of the reservoir to million gallons.

$$(43{,}560\ \text{ft}^3/\text{acre-ft})(22{,}000\ \text{acre-ft})(7.48\ \text{gal/ft}^3)(1\ \text{M}/1{,}000{,}000) = 7{,}168.2\ \text{mil gal}$$

The equation is:

$$\textbf{Copper sulfate, lb} = \frac{\textbf{(mil gal)(Dosage, mg/L)(8.34 lb/gal)}}{\textbf{\% Available Cu/100\%}}$$

$$\text{Copper sulfate, lb} = \frac{(7168.2\ \text{mil gal})(0.10\ \text{mg/L})(8.34\ \text{lb/gal})}{25\%/100\%} = 23{,}913\ \text{lb},$$

round to **24,000 lb of copper sulfate**

36. A treatment plant is adding 129.05 g/min of soda ash to its treated water. If the plant is producing water at a rate of 10.2 ft^3/s, what is the soda ash usage in pounds per day and dosage in milligrams per liter?

First, find the pounds per day of soda ash usage.

The equation is:

$$\textbf{Pounds per day, soda ash} = \textbf{(g/min)(1,440 min/d)(1 lb/454 g)}$$

$$\text{Pounds per day, soda ash} = (129.05\ \text{g/min})(1{,}440\ \text{min/d})(1\ \text{lb}/454\ \text{g})$$

Pounds per day, soda ash = 409.32 lb/d, round to **409 lb/d of soda ash**

Next, convert 10.2 ft^3/s to the number of million gallons per day.

$$\text{Number of mgd} = (10.2\ \text{ft}^3/\text{s})(86{,}400\ \text{s/d})(7.48\ \text{gal/ft}^3)(1\ \text{M}/1{,}000{,}000) = 6.59\ \text{mgd}$$

Then, rearrange the "pounds" equation to solve for dosage and calculate the dosage in milligrams per liter.

$$\text{Dosage, mg/L} = \frac{409.32\ \text{lb/d}}{(6.59\ \text{mgd})(8.34\ \text{lb/gal})} = \textbf{7.45 mg/L of soda ash}$$

37. A plant is treating water at a rate of 7.5 ft^3/s. If lime is being added at a rate of 90.8 g/min, what are the lime usage in pounds per day and the dosage in milligrams per liter?

First, find the pounds per day of lime usage.

The equation is:

$$\textbf{Pounds per day, lime} = \textbf{(g/min)(1,440 min/d)(1 lb/454 g)}$$

$$\text{Pounds per day, lime} = (90.8 \text{ g/min})(1{,}440 \text{ min/d})(1 \text{ lb/454 g}) = 288 \text{ lb/d of lime}$$

Next, convert 7.5 ft^3/s to the number of million gallons per day.

$$\begin{aligned}\text{Million gallons per day} &= (7.5 \text{ ft}^3\text{/s})(86{,}400 \text{ s/d})(7.48 \text{ gal/ft}^3)(1 \text{ M}/1{,}000{,}000)\\ &= 4.847 \text{ mgd}\end{aligned}$$

Rearrange and solve for dosage in milligrams per liter.

$$\text{Dosage, mg/L} = \frac{288 \text{ lb/d}}{(4.847 \text{ mgd})(8.34 \text{ lb/gal})} = \textbf{7.1 mg/L of lime}$$

38. A 5-min drawdown test shows that 421 mL of a cationic polymer are being injected. The specific gravity of the polymer is 1.09. If the plant is treating 10,850 gpm, what is the polymer dosage in milligrams per liter?

First, determine the number of milliliters per minute for the polymer.

$$\frac{421 \text{ mL}}{5 \text{ min}} = 84.2 \text{ mL/min}$$

Next, calculate the pounds per gallon for the polymer.

$$\textbf{Pounds per gallon} = \textbf{(Specific gravity)(8.34 lb/gal)}$$

$$\text{Pounds per gallon, polymer} = (1.09 \text{ sp gr})(8.34 \text{ lb/gal}) = 9.09 \text{ lb/gal}$$

Then, find the number of million gallons per day.

$$(10{,}850 \text{ gpm})(1{,}440 \text{ min/d})(1 \text{ M}/1{,}000{,}000) = 15.624 \text{ mgd}$$

The dosage equation is:

$$\textbf{Milligrams per liter} = \frac{\textbf{(mL/min)(1,440 min/d)(lb/gal, polymer)}}{\textbf{(3,785 mL/gal)(mgd)(8.34 lb/gal)}}$$

$$\text{mg/L, polymer} = \frac{(84.2 \text{ mL/min})(1{,}440 \text{ min/d})(9.09 \text{ lb/gal})}{(3{,}785 \text{ mL/gal})(15.624 \text{ mgd})(8.34 \text{ lb/gal})}$$

$$= \textbf{2.23 mg/L of polymer}$$

39. A water treatment plant is feeding 465 mL of alum per minute to treat 12 ft^3/s. The alum has a specific gravity of 1.27 with 48% purity. If the flow is increased to 15 ft^3/s, how many milliliters of alum will be used per minute if, at the same time, the dosage is increased by 1.5 ppm?

First, find the number of million gallons per day at a flow rate of 12 ft^3/s.

$$\textbf{Million gallons per day} = \frac{\textbf{(Number of ft}^3\textbf{/s)(7.48 gal/ft}^3\textbf{)(1,440 min/d)(60 s/min)}}{\textbf{1,000,000/M}}$$

$$\text{Million gallons per day} = \frac{(12\ \text{ft}^3/\text{s})(7.48\ \text{gal/ft}^3)(1{,}440\ \text{min/d})(60\ \text{s/min})}{1{,}000{,}000/\text{M}}$$

$$\text{Million gallons per day} = 7.755\ \text{mgd}$$

Next, find the number of pounds of alum used per day.

$$\textbf{Alum, lb/d} = \frac{\textbf{(Number of mL/min)(1,440 min/d)(Specific gravity)(\% Purity/100\%)}}{\textbf{(mL/gal)}}$$

Next, determine the number of pounds per gallon of alum used.

$$\textbf{Number of lb/gal} = \textbf{(Specific gravity)(8.34 lb/gal)}$$

$$\text{Number of lb/gal} = (1.27)(8.34\ \text{lb/gal}) = 10.592\ \text{lb/gal}$$

Now, find the number of pounds per day of alum using the "pounds" formula.

$$\text{Alum, lb/d} = \frac{(465\ \text{mL/min})(1{,}440\ \text{min/d})(10.592\ \text{lb/gal})(48\%\ \text{Purity})}{(3{,}785\ \text{mL/gal})(100\%)}$$

$$\text{Alum, lb/d} = 899.43\ \text{lb/d}$$

Next, substitute known values into the "pounds" formula.

$$\text{Dosage, ppm} = \frac{899.43\ \text{alum, lb/d}}{(7.755\ \text{mgd})(8.34\ \text{lb/gal})}$$

$$\text{Dosage, ppm} = 13.9\ \text{ppm}$$

$$\text{New dosage (increased)} = 13.9\ \text{ppm} + 1.5\ \text{ppm} = 15.4\ \text{ppm or } 15.4\ \text{mg/L}$$

Next, find the new flow in million gallons per day using the equation above.

$$\text{Million gallons per day} = \frac{(15\ \text{ft}^3/\text{s})(7.48\ \text{gal/ft}^3)(1{,}440\ \text{min/d})(60\ \text{s/min})}{1{,}000{,}000/\text{M}}$$

$$\text{Million gallons per day} = 9.694\ \text{mgd}$$

$$\text{Number of mL/min, alum} = \frac{(15.4\ \text{mg/L})(9.694\ \text{mgd})(3785\ \text{mL/gal})(8.34\ \text{lb/gal})(100\%)}{(1{,}440\ \text{min/d})(10.592\ \text{lb/gal})(48\%\ \text{Purity})}$$

Number of mL/min, alum = 643.68, round to **640 mL/min of alum**

40. A treatment plant is adding 70.26 g/min of soda ash to its treated water. If the plant is producing water at a rate of 8.4 ft³/s, what is the soda ash usage in pounds per day and dosage in milligrams per liter?

First, find the pounds per day of soda ash usage. The equation is:

Pounds per day, soda ash = (g/min)(1,440 min/d)(1 lb/454 g)

Pounds per day, soda ash = (70.26 g/min)(1,440 min/d)(1 lb/454 g)

Pounds per day, soda ash = 222.85 lb/d, round to 223 lb/d of soda ash

Next, convert 8 ft^3/s to the number of million gallons per day.

$$\text{Number of millions gallons per day} = (8.4\ \text{ft}^3/\text{s})(86{,}400\ \text{s/d})(7.48\ \text{gal/ft}^3)(1\ \text{M}/1{,}000{,}000) = 5.429\ \text{mgd}$$

Rearrange "pounds" equation and solve.

$$\text{Dosage, mg/L} = \frac{222.85\ \text{lb/d}}{(5.429\ \text{mgd})(8.34\ \text{lb/gal})} = \mathbf{4.9\ mg/L\ of\ soda\ ash}$$

41. A water treatment plant has a filter effluent flow of 6,945 gpm and is being treated with 684 gpd of a hypochlorite solution. If the desired dose is 1.5 mg/L, determine the concentration of the hypochlorite solution in percent.

First, convert 6,945 gpm filter flow to million gallons per day.

$$\frac{(6{,}945\ \text{gpm})\ (1{,}440\ \text{min})\ \ (1\ \text{mil gal})}{\text{day}\qquad 1{,}000{,}000\ \text{gal}} = 10\ \text{mgd}$$

Then, convert 684 gpd of the hypochlorite solution rate of flow to million gallons per day.

$$\frac{(684\ \text{gpd})\ \ (1\ \text{mil gal})}{1{,}000{,}000\ \text{gal}} = 0.000684\ \text{mgd}$$

Then, use the equal dosage equations.

$$(0.000684\ \text{mgd})(x\ \text{mg/L})(8.34\ \text{lb/gal}) = (10\ \text{mgd})(1.5\ \text{mg/L})(8.34\ \text{lb/gal})$$

$$x\ \text{mg/L} = \frac{(10\ \text{mgd})(1.5\ \text{mg/L})(8.34\ \text{lb/gal})}{(0.000684\ \text{mgd})(8.34\ \text{lb/gal})} = 21{,}930\ \text{mg/L}$$

Last, convert milligrams per liter into percent.

$$\frac{(21{,}930\ \text{mg/L})\ \ (1\%)}{10{,}000\ \text{mg/L}} = 2.193\%, \text{ round up to } \mathbf{2.2\%}$$

42. A plant is treating water at a rate of 13.7 ft^3/s. If lime is being added at a rate of 165.09 g/min, what is the lime usage in pounds per day and the dosage in milligrams per liter?

First, find the pounds per day of lime usage. The equation is:

$$\textbf{Pounds per day, lime} = \textbf{(g/min)(1,440 min/d)(1 lb/454 g)}$$

$$\text{Pounds per day, lime} = (165.09 \text{ g/min})(1{,}440 \text{ min/d})(1 \text{ lb}/454 \text{ g})$$

$$\text{Pounds per day, lime} = 523.63 \text{ lb/d, round to } \textbf{524 lb/d of lime}$$

Next, convert 13.7 ft^3/s to the number of million gallons per day.

$$\text{Number of million gallons per day} = (13.7 \text{ ft}^3\text{/s})(86{,}400 \text{ s/d})(7.48 \text{ gal/ft}^3)(1 \text{ M}/1{,}000{,}000) = 8.854 \text{ mgd}$$

Then, using the "pounds" equation, calculate the dosage in milligrams per liter.

$$\text{Dosage, mg/L} = \frac{523.63 \text{ lb/d}}{(8.854 \text{ mgd})(8.34 \text{ lb/gal})} = \textbf{7.09 mg/L of lime}$$

43. Determine the feed rate for alum in milliliters per minute under the following conditions:

Plant flow	**25.3 mgd**
Alum dosage rate	**7.25 mg/L**
Alum percentage	**49.0%**
Alum specific gravity	**1.24**

First, find the pounds per gallon for alum.

$$\text{lb/gal, alum} = (\text{Specific gravity})(8.34 \text{ lb/gal})$$

$$\text{lb/gal, alum} = (1.24)(8.34 \text{ lb/gal}) = 10.34 \text{ lb/gal}$$

The equation for dosage is:

$$\textbf{Dosage, mg/L} = \frac{\textbf{(mL/min)(1,440 min/d)(Alum, lb/gal)(\% Purity)}}{\textbf{(mgd)(8.34 lb/gal)(3,785 mL/gal)}}$$

Rearrange to solve for the unknown (milliliters per minute).

$$\text{Alum, mL/min} = \frac{(\text{Dosage, mg/L})(\text{mgd})(8.34 \text{ lb/gal})(3{,}785 \text{ ml/gal})}{(\text{Alum, lb/gal})(\% \text{ purity})(1{,}440 \text{ min/d})}$$

$$\text{Alum, mL/min} = \frac{(7.25 \text{ mg/L})(25.3 \text{ mgd})(8.34 \text{ lb/gal})(3{,}785 \text{ mL/gal})}{(10.34 \text{ lb/gal})(49.0\%/100\% \text{ purity})(1{,}440 \text{ min/d})}$$

$$= \textbf{794 mL/min of alum}$$

44. A water treatment plant has a flow of 4.8 mgd and is using 217.4 gpd of a hypochlorite solution. The chlorine dosage is 2.2 mg/L. Determine the concentration of the hypochlorite solution in percent.

First, convert the 217.4 gpd of hypochlorite solution to million gallons per day.

$$(217.4 \text{ gpd}) \frac{(1 \text{ mil gal})}{1{,}000{,}000 \text{ gal}} = 0.0002174 \text{ mgd}$$

Then, use the equal dosage equations.

$$(0.0002174 \text{ mgd})(x \text{ mg/L})(8.34 \text{ lb/gal}) = (480 \text{ mgd})(1.2 \text{ mg/L})(8.34 \text{ lb/gal})$$

$$\text{mg/L, hypochlorite} = \frac{(4.8 \text{ mgd})(2.2 \text{ mg/L})(8.34 \text{ lb/gal})}{(0.0002174 \text{ mgd})(8.34 \text{ lb/gal})} = 48{,}574 \text{ mg/L}$$

Last, convert milligrams per liter into percent.

$$\frac{(48{,}574 \text{ mg/L}) \quad (1\%)}{10{,}000 \text{ mg/L}} = 4.86\%, \text{ round to } \mathbf{4.9\%}$$

45. A treatment plant is adding 424.76 g/min of soda ash to its treated water. If the plant is producing water at a rate of 27 ft³/s, what is the soda ash usage in pounds per day and dosage in milligrams per liter?

First, find the pounds per day of soda ash usage. The equation is:

Pounds per day, soda ash = (g/min)(1,440 min/d)(1 lb/454 g)

Pounds per day, soda ash = (424.76 g/min)(1,440 min/d)(1 lb/454 g)

Pounds per day, soda ash = 1,347.26 lb/d, round to 1,350 lb/d of soda ash

Next, convert 27 ft^3/s to the number of million gallons per day.

$$\text{Number of million gallons per day} = (27 \text{ ft}^3\text{/s})(86{,}400 \text{ s/d})(7.48 \text{ gal/ft}^3)(1 \text{ M}/1{,}000{,}000) = 17.45 \text{ mgd}$$

Rearrange the "pounds" equation and solve.

$$\text{Dosage, mg/L} = \frac{1{,}347.26 \text{ lb/d}}{(17.45 \text{ mgd})(8.34 \text{ lb/gal})} = 9.26 \text{ mg/L, round to } \mathbf{9.3 \text{ mg/L of soda ash}}$$

46. Determine the feed rate for alum in milliliters per minute under the following conditions:

Plant flow	**21.5 mgd**
Alum dosage rate	**8.30 mg/L**
Alum percentage	**48.0%**
Alum specific gravity	**1.29**

First, find the pounds per gallon for alum.

$$\text{lb/gal, alum} = (\text{Specific gravity})(8.34\ \text{lb/gal})$$

$$\text{lb/gal, alum} = (1.29)(8.34\ \text{lb/gal}) = 10.759\ \text{lb/gal}$$

The equation for dosage is:

$$\textbf{Dosage, mg/L} = \frac{\textbf{(mL/min)(1,440 min/d)(Alum, lb/d)(\% purity)}}{\textbf{(mgd)(8.34 lb/gal)(3,785 mL/gal)}}$$

Rearrange to solve for the unknown (milliliters per minute).

$$\text{Alum, mL/min} = \frac{(\text{Dosage, mg/L})(\text{mgd})(8.34\ \text{lb/gal})(3{,}785\ \text{mL/gal})}{(\text{Alum, lb/gal})(\%\ \text{purity})(1{,}440\ \text{min/d})}$$

$$\text{Alum, mL/min} = \frac{(8.30\ \text{mg/L})(21.5\ \text{mgd})(8.34\ \text{lb/gal})(3{,}785\ \text{mL/gal})}{(10.759\ \text{lb/gal})(48.0\%/100\%\ \text{purity})(1{,}440\ \text{min/d})}$$

$$= \textbf{757 mL/min of alum}$$

47. How many fluid ounces of sodium hypochlorite (5.7% available chlorine) are required to disinfect a well with the following parameters: (1) depth is 316 ft; (2) 14-in. diameter well casing extends down to 100.0 ft; (3) the remainder is a 10.0-in. diameter casing; (4) the residual desired dose is 50.0 mg/L; (5) the depth to water is 88 ft; and (6) the chlorine demand is 14 mg/L. Assume the sodium hypochlorite solution weighs 9.03 lb/gal.

First, find the diameter, in feet, for both well casings.

$$\text{Diameter for 14-in. casing} = \frac{(14\ \text{in.})(1\ \text{ft})}{12\ \text{in.}} = 1.167\ \text{ft}$$

$$\text{Diameter for 10.0-in. casing} = \frac{(10.0\ \text{in.})(1\ \text{ft})}{12\ \text{in.}} = 0.833\ \text{ft}$$

(continued)

Then, find the length, in feet, of water in the casing.

Length of water-filled casing = Depth of well – Depth of water to top of casing

Length of water-filled casing for 14-in. diameter = 100.0 ft – 88 ft = 12 ft

Length of water-filled casing for 10.0-in. diameter = 316 ft – 100 ft = 216 ft

Next, determine the volume of water, in gallons, in the well casings using the following formula:

Volume, in gal = **(0.785)(Diameter)^2(Length)(7.48 gal/ft^3)**

Volume, gal for 14-in. casing = (0.785)(1.167 ft)(1.167 ft)(12 ft)(7.48 gal/ft^3)
= 96 gal

Volume, gal for 10.0-in. casing = (0.785)(0.833 ft)(0.833 ft)(216 ft)(7.48 gal/ft^3)
= 880 gal

Total water in well casings = 96 gal + 888 gal = 976 gal

Next, determine the number of million gallons.

Million gallons = (976 gal)(1 M/1,000,000) = 0.000976 mil gal

Then, find the chlorine required.

Chlorine required = Chlorine demand + Chlorine residual

Cl_2 req. = 14 mg/L + 50 mg/L = 64 mg/L

Then, using the "pounds" equation, calculate the number of pounds of sodium hypochlorite.

$$\text{Pounds, sodium hypochlorite} = \frac{(0.000976 \text{ mil gal})(64 \text{ mg/L})(8.34 \text{ lb/gal})}{5.7\% \text{ Available chlorine}/100\%}$$

Pounds, sodium hypochlorite = 9.139 lb

Next, calculate the number of gallons of sodium hypochlorite required.

(9.139 lb)/(9.03 lb/gal) = 1.012 gal

Last, convert to fluid ounces.

1.012 gal(128 oz/gal) = 129.54 oz, round to **130 oz of sodium hypochlorite**

48. Determine the feed rate for alum in milliliters per minute under the following conditions:

Plant flow	**17.45 mgd**
Alum dosage rate	**18.0 mg/L**
Alum percentage	**47.8%**
Alum specific gravity	**1.23**

First, find the pounds per gallon for alum.

$$\text{Pounds per gallon, alum} = (\text{Specific gravity})(8.34 \text{ lb/gal})$$

$$\text{Pounds per gallon, alum} = (1.23)(8.34 \text{ lb/gal}) = 10.258 \text{ lb/gal}$$

The equation for dosage is:

$$\textbf{Dosage, mg/L} = \frac{\textbf{(mL/min)(1,440 min/d)(Alum, lb/gal)(\% Purity)}}{\textbf{(mgd)(8.34 lb/gal)(3,785 mL/gal)}}$$

Rearrange to solve for the unknown (milliliters per minute).

$$\text{Alum, mL/min} = \frac{(\text{Dosage, mg/L})(\text{mgd})(8.34 \text{ lb/gal})(3{,}785 \text{ mL/gal})}{(\text{Alum, lb/gal})(\% \text{ purity})(1{,}440 \text{ min/d})}$$

$$\text{Alum, mL/min} = \frac{(18.0 \text{ mg/L})(17.45 \text{ mgd})(8.34 \text{ lb/gal})(3{,}785 \text{ mL/gal})}{(10.258 \text{ lb/gal})(47.8\%/100\% \text{ purity})(1{,}440 \text{ min/d})}$$

Alum, mL/min = 1,404 mL/min, round to **1,400 mL/min of alum**

LIME DOSAGE PROBLEMS

49. Using jar tests, an operator has determined that 21.4 mg/L of alum are required for optimum performance in treating the raw water at his treatment plant. What is the lime dose required if the raw water alkalinity is 14.5 mg/L? Assume the residual alkalinity required for complete precipitation is 15 mg/L.

The following applies:

1 mg/L of alum reacts with 0.45 mg/L of alkalinity

1 mg/L of alum reacts with 0.35 mg/L of lime

First, determine the quantity of alkalinity that will react with the alum.

The equation is:

$$\frac{\textbf{0.45 mg/L alkalinity}}{\textbf{1 mg/L alum}} = \frac{\boldsymbol{x}\textbf{ mg/L alkalinity}}{\textbf{Number of mg/L alum required}}$$

(continued)

Substituting:

$$\frac{0.45 \text{ mg/L alkalinity}}{1 \text{ mg/L alum}} = \frac{x \text{ mg/L alkalinity}}{21.4 \text{ mg/L alum}}$$

$$x \text{ mg/L alkalinity} = \frac{(0.45 \text{ mg/L alkalinity})(21.4 \text{ mg/L alum})}{1 \text{ mg/L alum}}$$

$$x \text{ mg/L alkalinity} = 9.63 \text{ mg/L}$$

Next, calculate the total alkalinity required.

The equation is:

Total alkalinity required, mg/L = Alkalinity reacting with alum, mg/L + Residual alkalinity for precipitation, mg/L

$$\text{Total alkalinity, mg/L} = 9.63 \text{ mg/L} + 15 \text{ mg/L}$$

$$\text{Total alkalinity, mg/L} = 24.63 \text{ mg/L}$$

Now, determine the amount of alkalinity that must be added to the raw water.

The equation is:

Alkalinity needed for raw water, mg/L = Total alkalinity required, mg/L – Alkalinity in raw water, mg/L

$$\text{Alkalinity needed, mg/L} = 24.63 \text{ mg/L} - 14.5 \text{ mg/L}$$

$$\text{Alkalinity needed} = 10.13 \text{ mg/L}$$

Last, determine the lime dosage.

The equation is:

$$\frac{\textbf{0.35 mg/L lime}}{\textbf{0.45 mg/L alkalinity}} = \frac{\textbf{Number of mg/L lime}}{\textbf{Alkalinity needed, mg/L}}$$

Substituting:

$$\frac{0.35 \text{ mg/L lime}}{0.45 \text{ mg/L alkalinity}} = \frac{\text{Number of mg/L lime}}{10.13 \text{ mg/L alkalinity}}$$

$$\text{Number of mg/L lime} = \frac{(0.35 \text{ mg/L lime})(10.13 \text{ mg/L, alkalinity})}{0.45 \text{ mg/L alkalinity}}$$

Number of mg/L lime = 7.88 mg/L, round to **7.9 mg/L of lime**

50. Given the following data, calculate the dosage of lime needed for a conventional water treatment plant using alum:

Alum needed based on jar testing = 38.7 mg/L.

Raw water alkalinity = 32.6 mg/L.

Assume the residual alkalinity that is required for complete precipitation is 20 mg/L.

1 mg/L of alum reacts with 0.45 mg/L of alkalinity.

1 mg/L of alum reacts with 0.35 mg/L of lime.

First, determine the quantity of alkalinity that will react with the alum.

The equation is:

$$\frac{\textbf{0.45 mg/L alkalinity}}{\textbf{1 mg/L alum}} = \frac{\boldsymbol{x}\textbf{ mg/L alkalinity}}{\textbf{Number of mg/L alum required}}$$

Substituting:

$$\frac{0.45 \text{ mg/L alkalinity}}{1 \text{ mg/L alum}} = \frac{x \text{ mg/L alkalinity}}{38.7 \text{ mg/L alum}}$$

$$x \text{ mg/L alkalinity} = \frac{(0.45 \text{ mg/L alkalinity})(38.7 \text{ mg/L alum})}{1 \text{ mg/L alum}}$$

$$x \text{ mg/L alkalinity} = 17.415 \text{ mg/L}$$

Next, calculate the total alkalinity required.

The equation is:

Total alkalinity required, mg/L = Alkalinity reacting with alum, mg/L + Residual alkalinity for precipitation, mg/L

$$\text{Total alkalinity, mg/L} = 17.415 \text{ mg/L} + 20 \text{ mg/L}$$

$$\text{Total alkalinity, mg/L} = 37.415 \text{ mg/L}$$

Now, determine the amount of alkalinity that must be added to the raw water.

The equation is:

Alkalinity needed for raw water, mg/L = Total alkalinity required, mg/L – Alkalinity in raw water, mg/L

$$\text{Alkalinity needed, mg/L} = 37.415 \text{ mg/L} - 32.6 \text{ mg/L}$$

$$\text{Alkalinity needed} = 4.815 \text{ mg/L}$$

(continued)

Last, determine the lime dosage.

The equation is:

$$\frac{\textbf{0.35 mg/L lime}}{\textbf{0.45 mg/L alkalinity}} = \frac{\textbf{Number of mg/L lime}}{\textbf{Alkalinity needed, mg/L}}$$

Substituting:

$$\frac{0.35 \text{ mg/L lime}}{0.45 \text{ mg/L alkalinity}} = \frac{\text{Number of mg/L lime}}{4.815 \text{ mg/L alkalinity}}$$

$$\text{Number of mg/L lime} = \frac{(0.35 \text{ mg/L lime})(4.815 \text{ mg/L, alkalinity})}{0.45 \text{ mg/L alkalinity}}$$

Number of mg/L lime = 3.745 mg/L, round to **3.7 mg/L of lime**

SOLUTION FEED RATE CALCULATIONS

51. Calculate the feed rate for fluorosilicic acid in gallons per day given the following data:

Flow rate is 8.3 mgd
Fluoride (F) desired is 1.20 mg/L
Fluoride in raw water is 0.22 mg/L
Treated with 20.0% solution of H_2SiF_6
Fluoride ion percent is 79%
H_2SiF_6 weighs 9.8 lb/gal

F req. = F desired – F in raw water

F req. = 1.20 mg/L – 0.22 mg/L = 0.98 mg/L

$$\frac{(\% \text{ solution})(10{,}000 \text{ mg/L})}{1\%} = \text{Number of mg/L}$$

(20.0% solution of H_2SiF_6)(10,000 mg/L per 1%) = 200,000 mg/L

The equation is:

(F, mg/L)(mgd)(8.34 lb/gal) = (% as mg/L)(mgd flow)(Solution, lb/gal)(% F ion)

(0.98 mg/L F)(8.3 mgd)(8.34 lb/gal) = (200,000 mg/L)(x mgd flow)(9.8 lb/gal)(79%/100% F)

Solve for x.

$$x \text{ mgd flow} = \frac{(0.98 \text{ mg/L F})(8.3 \text{ mgd})(8.34 \text{ lb/gal})}{(200{,}000 \text{ mg/L})(9.8 \text{ lb/gal})(79\%/100\% \text{ F})} = 0.0000438 \text{ mgd}$$

Convert million gallons per day to gallons.

$$(0.0000438 \text{ mgd})(1{,}000{,}000 \text{ gal/1 M}) = 43.8 \text{ gpd, round to } \mathbf{44\ gpd\ H_2SiF_6}$$

52. Calculate the feed rate for fluorosilicic acid in milliliters per minute given the following data:

Flow rate is 22.6 mgd | **Treated with 25% solution of H_2SiF_6**
Fluoride (F) desired is 1.20 mg/L | **Fluoride ion percent is 79.2%**
Fluoride in raw water is 0.60 mg/L | **H_2SiF_6 weighs 9.8 lb/gal**

$$\mathbf{F\ required = F\ desired - F\ in\ raw\ water}$$

$$\text{F req.} = 1.20 \text{ mg/L} - 0.60 \text{ mg/L} = 0.60 \text{ mg/L}$$

$$\frac{(\%\text{ solution})(10{,}000 \text{ mg/L})}{1\%} = \text{Number of mg/L}$$

$$(25\% \text{ solution of } H_2SiF_6)(10{,}000 \text{ mg/L per } 1\%) = 250{,}000 \text{ mg/L}$$

The equation is:

$$\mathbf{(F, mg/L)(mgd)(8.34\ lb/gal) = (\%\ as\ mg/L)(mgd\ flow)(Solution, lb/gal)(\%\ F\ ion)}$$

$$(0.6 \text{ mg/L F})(22.6 \text{ mgd})(8.34 \text{ lb/gal})$$

$$= (250{,}000 \text{ mg/L})(x \text{ mgd flow})(9.8 \text{ lb/gal})(79.2\%/100\% \text{ F})$$

Solve for x.

$$x \text{ mgd flow} = \frac{(0.60 \text{ mg/L F})(22.6 \text{ mgd})(8.34 \text{ lb/gal})}{(250{,}000 \text{ mg/L})(9.8 \text{ lb/gal})(79.2\%/100\% \text{ F})} = 0.0000582 \text{ mgd}$$

Convert million gallons per day to gallons.

$$(0.0000582 \text{ mgd})(1{,}000{,}000 \text{ gal/1 M}) = 58.2 \text{ gpd } H_2SiF_6$$

Then, convert gallons per day to milliliters per minute.

$$\text{mL/min, } H_2SiF_6 = (58.2 \text{ gpd})(3{,}785 \text{ mL/gal})(1 \text{ d}/1{,}440 \text{ min})$$

$$\text{mL/min, } H_2SiF_6 = 152.98 \text{ mL/min, round to } \mathbf{150\ mL/min\ of\ H_2SiF_6}$$

53. Calculate the feed rate for fluorosilicic acid in gallons per day given the following data:

Flow rate is 11.4 mgd	**Treated with 20% solution of H_2SiF_6**
Fluoride (F) desired is 1.20 mg/L	**Fluoride ion percent is 79%**
Fluoride in raw water is 0.50 mg/L	**H_2SiF_6 weighs 9.8 lb/gal**

F req. = F desired – F in raw water

F req. = 1.20 mg/L – 0.50 mg/L = 0.70 mg/L

$$\frac{(\%\ \text{solution})(10{,}000\ \text{mg/L})}{1\%} = \text{Number of mg/L}$$

20% solution of H_2SiF_6 = 200,000 mg/L

The equation is:

(F, mg/L)(mgd)(8.34 lb/gal) = (% as mg/L)(mgd flow)(Solution, lb/gal)(% F ion)

(0.7 mg/L F)(11.4 mgd)(8.34 lb/gal) = (200,000 mg/L)(x mgd flow)(9.8 lb/gal)(0.79 F)

Solve for x.

$$x\ \text{mgd flow} = \frac{(0.70\ \text{mg/L F})(11.4\ \text{mgd})(8.34\ \text{lb/gal})}{(200{,}000\ \text{mg/L})(9.8\ \text{lb/gal})(79\%/100\%\ \text{F})} = 0.00004298\ \text{mgd}$$

Convert million gallons per day to gallons.

(0.00004298 mgd)(1,000,000 gal/1 M) = 42.98 gpd, round to **43 gpd H_2SiF_6**

54. Calculate the feed rate for fluorosilicic acid in milliliters per minute given the following data:

Flow rate is 92 mgd	**Treated with 25% solution of H_2SiF_6**
Fluoride (F) desired is 1.27 mg/L	**Fluoride ion percent is 79.1%**
Fluoride in raw water is 0.27 mg/L	**H_2SiF_6 weighs 9.8 lb/gal**

F req. = F desired – F in raw water

F req. = 1.27 mg/L – 0.27 mg/L = 1.00 mg/L

$$\frac{(\%\ \text{solution})(10{,}000\ \text{mg/L})}{1\%} = \text{Number of mg/L}$$

25.0% solution of H_2SiF_6 = 250,000 mg/L

The equation is:

(F, mg/L)(mgd)(8.34 lb/gal) = (% as mg/L)(mgd flow)(Solution, lb/gal)(% F ion)

(1.0 mg/L F)(92 mgd)(8.34 lb/gal) = (250,000 mg/L)(x mgd flow)(9.8 lb/gal)(79.1%/100% F)

Solve for x.

$$x \text{ mgd flow} = \frac{(1.0 \text{ mg/L F})(92 \text{ mgd})(8.34 \text{ lb/gal})}{(250{,}000 \text{ mg/L})(9.8 \text{ lb/gal})(79.1\%/100\% \text{ F})} = 0.000396 \text{ mgd}$$

Convert millions gallons per day to gallons.

$$(0.000396 \text{ mgd})(1{,}000{,}000 \text{ gal/1 M}) = 396 \text{ gpd } H_2SiF_6$$

Then, convert gallons per day to milliliters per minute.

$$\text{mL/min, } H_2SiF_6 = (396 \text{ gpd})(3{,}785 \text{ mL/gal})(1 \text{ d/1,440 min})$$

$$\text{mL/min, } H_2SiF_6 = 1{,}041 \text{ mL/min, round to } \mathbf{1{,}000 \text{ mL/min of } H_2SiF_6}$$

POUNDS REMOVAL PROBLEMS

55. Find the amount of iron and manganese removed per year from a plant that treats an average of 12.0 mgd if the average iron concentration is 1.14 ppm and the average manganese concentration is 0.14 ppm. The removal efficiency is 80.7% for the iron and 68.8% for the manganese.

First, calculate the iron removal in parts per million.

$$(1.14 \text{ ppm})(80.7\%/100\%) = (1.14 \text{ ppm})(0.807) = 0.920 \text{ ppm or } 0.920 \text{ mg/L}$$

Do the same for the manganese.

$$(0.14 \text{ ppm})(68.8\%/100\%) = (0.14 \text{ ppm})(0.688) = 0.0963 \text{ ppm or } 0.0963 \text{ mg/L}$$

Determine the amount of water produced, in million gallons, for the year.

$$(12.0 \text{ mgd})(365 \text{ d/yr}) = 4{,}380 \text{ mil gal/yr}$$

Next, using the "pounds" equation, solve for the number of pounds per year for iron and manganese.

Pounds per year = (mil gal/yr)(Dosage, mg/L)(8.34 lb/gal)

Pounds per year, Fe removal = (4,380 mil gal/yr)(0.920 mg/L)(8.34 lb/gal)

= 33,607 lb/yr, round to **33,600 lb/yr of Fe removed**

Pounds per year, Mn removal = (4,380 mil gal/yr)(0.0963 mg/L)(8.34 lb/gal)

= 3,518 lb/yr, round to **3,500 lb/yr of Mn removed**

56. Find the amount of iron and manganese removed per year from a plant that treats an average of 32 mgd if the average iron concentration is 2.18 mg/L and the average manganese concentration is 0.34 mg/L. The removal efficiency is 89.2% for the iron and 72.2% for the manganese.

First, calculate the iron removal in milligrams per liter.

$$(2.18 \text{ mg/L})(89.2\%/100\%) = 2.18 \text{ mg/L}(0.892) = 1.945 \text{ mg/L}$$

Now for manganese.

$$(0.34 \text{ mg/L})(72.2\%/100\%) = 0.34 \text{ mg/L}(0.722) = 0.245 \text{ mg/L}$$

Determine the amount of water produced, in million gallons, for the year.

$$(32 \text{ mgd})(365 \text{ d/yr}) = 11{,}680 \text{ mil gal/yr}$$

Next, using the "pounds" equation, solve for the number of pound per year for iron and manganese.

Pounds per year = (mil gal/yr)(Dosage, mg/L)(8.34 lb/gal)

Pounds per year, Fe removal = (11,680 mil gal/yr)(1.945 mg/L)(8.34 lb/gal)

= 189,465 lb/yr, round to **190,000 lb/yr of Fe removed**

Pounds per year, Mn removal = (11,680 mil gal/yr)(0.245 mg/L)(8.34 lb/gal)

= 23,866 lb/yr, round to **24,000 lb/yr of Mn removed**

57. Find the amount of iron and manganese removed per year from a plant that treats an average of 75.5 mgd if the average iron concentration is 1.76 ppm and the average manganese concentration is 0.29 ppm. The removal efficiency is 87.4% for the iron and 71.3% for the manganese.

First, calculate the iron removal in parts per million.

$$(1.76 \text{ ppm})(87.4\%/100\%) = 1.76 \text{ ppm}(0.874) = 1.538 \text{ ppm or } 1.538 \text{ mg/L}$$

Now for manganese.

$$(0.29 \text{ ppm})(71.3\%/100\%) = 0.29 \text{ ppm}(0.713) = 0.207 \text{ ppm or } 0.207 \text{ mg/L}$$

Determine the amount of water produced, in million gallons, for the year.

$$(75.5 \text{ mgd})(365 \text{ d/yr}) = 27{,}557.5 \text{ mil gal/yr}$$

Next, using the "pounds" equation, solve for the number of pounds per year for iron and manganese.

Pounds per year = (mil gal/yr)(Dosage, mg/L)(8.34 lb/gal)

Pounds per year, Fe removal = (27,557.5 mil gal/yr)(1.538 mg/L)(8.34 lb/gal)

= 353,478 lb/yr, round to **353,000 lb/yr of Fe removed**

Pounds per year, Mn removal = (27,557.5 mil gal/yr)(0.207 mg/L)(8.34 lb/gal)

= 47,575 lb/yr, round to **48,000 lb/yr of Mn removed**

58. A water district is treating 1,250 mgd of sea water. The sea water contains 2,840 mg/L total salts. How many pounds per day and pounds per year of salts are removed if 99.1% efficiency is achieved?

The equation is:

Pounds per day = (mgd)(Concentration, mg/L)(8.34 lb/gal)(Removal efficiency)

Pounds per day of salts removed = (1,250)(2,840 mg/L)(8.34 lb/gal)(99.1%/100%)

= **29,340,537 lb/d**

Then, find the pounds of salt per year removed.

Pounds per day of salt removed = (29,340,537 lb/d)(365 days/yr)

= 10,709,296,010 lb/yr,
round to **10,700,000,000 lb/yr or 10.7 billion lb/yr**

PERCENT SOLUTION AND MIXTURE PROBLEMS

59. How many gallons of a 9.2% solution must be mixed with a 3.6% solution to make exactly 250 gal of a 5.0% solution?

There are two ways to solve dilution problems. The dilution triangle is perhaps the easiest, and is shown below for the next three problems.

Solve the problem using the dilution triangle. The two numbers on the left are the existing concentrations of 9.2% and 3.6%. The number in the center, 5.0%, is the desired concentration. The numbers on the right are determined by subtracting diagonally the existing concentrations from the desired concentration.

9.2%		1.4*	1.4 parts of the 9.2% solution are required for every 5.6 parts
	5.0%		
3.6%		4.2†	4.2 parts of the 3.6% solution are required for every 5.6 parts
		5.6 total parts	

$$\frac{1.4 \text{ parts } (250 \text{ gal})}{5.6 \text{ parts}} = 62.5 \text{ gal, round to } \mathbf{62 \text{ gal of the } 9.2\% \text{ solution}}$$

$$\frac{4.2 \text{ parts } (250 \text{ gal})}{5.6 \text{ parts}} = \frac{187.5 \text{ gal, round to } \mathbf{188^{\ddagger} \text{ gal of the } 3.6\% \text{ solution}}}{250 \text{ gal (added here to cross-check math)}}$$

To make the 250 gal of the 5.0% solution, mix 62 gal of the 9.2% solution with 188 gal of the 3.6% solution.

* 1.4 is determined by subtracting diagonally 3.6% from 5.0%. The negative sign is dropped.

† 4.2 is determined by subtracting diagonally 9.2% from 5.0%.

‡ Should actually round to 190 gal, but this would make 252 gal in the field. So for practical reasons when mixing in the field use 188 gal.

60. An operator needs to prepare a solution containing 125 gal of 3.0% hypochlorite. How many gallons of a 6.7% solution must be mixed with a 1.0% solution to make the required solution?

Solve the problem using the dilution triangle.

6.7%   2.0   2.0 parts of the 6.7% solution are required for every 5.7 parts

3.0%

1.0%   3.7   3.7 parts of the 1% solution are required for every 5.7 parts

5.7 total parts

$$\frac{2.0 \text{ parts}(125 \text{ gal})}{5.7 \text{ parts}} = 43.9 \text{ gal, round to } \mathbf{44 \text{ gal of the } 6.7\% \text{ solution}}$$

$$\frac{3.7 \text{ parts}(125 \text{ gal})}{5.7 \text{ parts}} = \frac{81.1}{125 \text{ gal}} \text{ gal, round to } \mathbf{81 \text{ gal of the } 1\% \text{ solution}}$$

To make the 125 gal of the 3.0% solution, mix 44 gal of the 6.7% solution with 81 gal of the 1% solution.

61. An operator needs to prepare 50 gal of a 4.3% solution of hypochlorite. The chlorine room contains 14 gal of a 15% solution and 110 gal of a 2.3% solution. What would you do to make the desired solution if you don't want to use the 2.3% solution?

First, determine if you can simply mix some of the 15% solution with water.

15%   4.3   4.3 parts of the 15% solution are required for every 15 parts

4.3%

0%   10.7   10.7 parts of the water are needed for every 15 parts

15 total parts

But, is there enough of the 15% solution to make 50 gal of a 4.3% solution?

$$\frac{4.3 \text{ parts } (50 \text{ gal})}{15 \text{ parts}} = 14.3 \text{ gal of the } 15\% \text{ solution is needed.}$$

This is more than what you have. The 2.3% solution will be needed. Use the dilution triangle again.

15%   2.0   2.0 parts of the 15% solution are required for every 12.7 parts

4.3%

2.3%   10.7   10.7 parts of the 2.3% solution are required for every 12.7 parts

12.7 total parts

(continued)

$$\frac{2.0 \text{ parts } (50 \text{ gal})}{12.7 \text{ parts}} = 7.9 \text{ gal, round to } \mathbf{8 \text{ gal of the } 15\% \text{ solution}}$$

$$\frac{10.7 \text{ parts } (50 \text{ gal})}{12.7 \text{ parts}} = \frac{42.1}{50 \text{ gal}} \text{ gal, round to } \mathbf{42 \text{ gal of the } 2.3\% \text{ solution}}$$

To make 50 gal of the 4.3% solution, mix 8 gal of the 15% solution with 42 gal of the 2.3% solution.

62. What is the percent strength of a solution if 240 gal of a 4.0% solution are mixed with 760 gal of a 15.8% solution? The density of both solutions is 8.34 lb/gal.

The equation is:

$$\textbf{Percent mixture strength} = \frac{\textbf{Solution 1 gal(Available \%/100\%) + Solution 2 gal(Available \%/100\%)}}{\textbf{Gallons of solution 1 + Gallons of solution 2}} \times \mathbf{100\%}$$

$$\text{Percent mixture strength} = \frac{240 \text{ gal}(4.0\%/100\%) + 760 \text{ gal}(15.8\%/100\%)}{240 \text{ gal} + 760 \text{ gal}} \times 100\%$$

$$\text{Percent mixture strength} = \frac{9.6 \text{ gal} + 120.08 \text{ gal}}{1{,}000 \text{ gal}} \times 100\% = \frac{129.68 \text{ gal}}{1{,}000 \text{ gal}} \times 100\%$$

$$= \mathbf{13\% \text{ strength}}$$

63. What percent hypochlorite solution would result if 200 gal of a 15% solution were mixed with 120 gal of a 4.0% solution? Assume both solutions have the same density and are measured to the nearest gallon (thus they both have three significant figures).

First, find the total volume that would result from mixing these two solutions.

$$\text{Total volume} = 200 \text{ gal} + 120 \text{ gal} = 320 \text{ gal}$$

Another way to solve percent solution problems or to find volumes is to use the following equation:

$$\textbf{(Concentration}_1\textbf{)(Volume}_1\textbf{)} + \textbf{(Concentration}_2\textbf{)(Volume}_2\textbf{)}$$
$$= \textbf{(Concentration}_3\textbf{)(Volume}_3\textbf{)},$$

or abbreviate as:

$$C_1V_1 + C_2V_2 = C_3V_3$$

where C_1 and C_2 = percent concentration of the two solutions before being mixed
V_1 and V_2 = volume of the two solutions before being mixed
C_3 and V_3 = the resulting percent concentration and volume, respectively

Substitute and divide all concentrations and volumes by 100%.

$$\frac{(15\%)(200 \text{ gal})}{100\%} + \frac{(4.0\%)(120 \text{ gal})}{100\%} = \frac{C_3(320 \text{ gal})}{100\%}$$

Then, reduce by cancelling out percent on left side of equation.

$$30 \text{ gal} + 4.8 \text{ gal} = \frac{C_3(320 \text{ gal})}{100\%}$$

$$34.8 \text{ gal} = \frac{C_3(320 \text{ gal})}{100\%}$$

Solve for C_3.

$$C_3 = \frac{(34.8 \text{ gal})(100\%)}{320 \text{ gal}}$$

$C_3 = 10.875$, round to **11% final solution**

64. A solution of lime needs to be prepared for a jar test. How many grams of quicklime, CaO, would you mix with 1 L of water to make a 0.1% (weight-volume) solution?

The equation for weight-volume percent is:

$$\textbf{Percent of solution} = \frac{\textbf{Weight of solute, grams}}{\textbf{Volume of solution, mL}} \times \textbf{100\%}$$

First, determine the volume in milliliters. The following applies: 1 L = 1,000 mL.

$$0.1\% = \frac{\text{Weight of solute, g}}{1{,}000 \text{ mL}} \times 100\%$$

Solve for weight of solute, in grams, by multiplying both side of the equation by 1,000 mL and dividing both sides by 100%.

$$\text{Weight of solute, g} = \frac{(0.1\%)(1{,}000 \text{ mL})}{100\%} = \textbf{1 g of CaO dissolved in 1 L of water}$$

65. A storage tank contains 345 gal of a 14% zinc orthophosphate (ZOP) solution (soln.) that weighs 10.47 lb/gal. Note that 1,000 gal of a 11.8% ZOP solution are added to the tank. Paperwork on the new ZOP indicates that the specific gravity is 1.16 lb/gal. What is the percent strength of the resulting solution?

First, determine the weight in pounds of the 11.8% solution.

Weight of 11.8% solution = (Specific gravity)(8.34 lb/gal)

Weight of 11.8% solution (solution) = (1.16 sp gr)(8.34 lb/gal) = 9.67 lb/gal

The equation is:

$$\textbf{Percent mixture strength} = \frac{\textbf{Soln. 1, gal(lb/gal)(Avail. ZOP\%/100\%) + Soln. 2, gal(lb/gal)(Avail. ZOP\%/100\%)}}{\textbf{Soln. 1, gal(lb/gal) + Soln. 2, gal(lb/gal)}} \times \textbf{100\%}$$

$$\text{Percent mixture strength} = \frac{345 \text{ gal}(10.47 \text{ lb/gal})(14\%/100\%) + 1{,}000 \text{ gal}(9.67 \text{ lb/gal})(11.8\%/100\%)}{345 \text{ gal}(10.47 \text{ lb/gal}) + 1{,}000 \text{ gal}(9.67 \text{ lb/gal})} \times 100\%$$

Percent mixture strength =

$$\frac{505.7 \text{ lb} + 1{,}141.06 \text{ lb}}{3{,}612.15 \text{ lb} + 9{,}670 \text{ lb}} \times 100\% = \frac{1{,}646.76 \text{ lb}}{13{,}282.15 \text{ lb}} \times 100\%$$

= **12% mixture strength of ZOP**

66. What percent polymer solution would result if 1,400 gal of a 12% solution were mixed with 440 gal of a 5.5% solution? Assume both solutions have the same density.

First, find the total volume that would result from mixing these two solutions.

Total volume = 1,400 gal + 440 gal = 1,840 gal

The equation is:

$$\mathbf{C_1V_1 + C_2V_2 = C_3V_3}$$

where C_1 and C_2 = percent concentration of the two solutions before being mixed
V_1 and V_2 = volume of the two solutions before being mixed
C_3 and V_3 = the resulting percent concentration and volume, respectively

Substitute and divide all concentrations and volumes by 100%.

$$\frac{(12\%)(1{,}400 \text{ gal})}{100\%} + \frac{(5.5\%)(440 \text{ gal})}{100\%} = \frac{C_3(1{,}840 \text{ gal})}{100\%}$$

Then, reduce left side of equation by dividing by 100%.

$$168 \text{ gal} + 24.2 \text{ gal} = \frac{C_3(1{,}840 \text{ gal})}{100\%}$$

Solve for C_3.

$$C_3 = \frac{(192.2 \text{ gal})(100\%)}{1{,}840 \text{ gal}} = \frac{(192.2 \text{ gal})}{1{,}840 \text{ gal}}(100\%)$$

C_3 = 10.45, round to **10% final solution**

67. A 4.0% hypochlorite solution is required. If exactly 475 gal are needed, how many gallons of a 9.4% solution must be mixed with a 2.5% solution to make the required solution?

Solve the problem using the dilution triangle.

9.4%		1.5	1.5 parts of the 9.4% solution are required for every 6.9 parts
	4.0%		
2.5%		5.4	5.4 parts of the 2.5% solution are required for every 6.9 parts
		6.9 total parts	

$$\frac{1.5 \text{ parts}(475 \text{ gal})}{6.9 \text{ parts}} = \textbf{103 gal of the 9.4\% solution}$$

$$\frac{5.4 \text{ parts } (475 \text{ gal})}{6.9 \text{ parts}} = \textbf{372 gal of the 2.5\% solution}$$

475 gal

Mix 103 gal of the 9.4% solution with 372 gal of the 2.5% solution to get the final solution of 4.0%.

68. How many gallons of a 23.4% solution must be mixed with an 8.4% solution to make exactly 500 gal of a 15% solution?

Solve the problem using the dilution triangle.

23.4%		6.6	6.6 parts of the 23.4% solution are required for every 15 parts
	15%		
8.4%		8.4	8.4 parts of the 8.4% solution are required for every 15 parts
		15 total parts	

$$\frac{6.6 \text{ parts}(500 \text{ gal})}{15 \text{ parts}} = \textbf{220 gal of the 23.4\% solution}$$

$$\frac{8.4 \text{ parts } (500 \text{ gal})}{15 \text{ parts}} = \textbf{280 gal of the 8.4\% solution}$$

500 gal

To make the 500 gal of the 15% solution, mix 220 gal of the 23.4% solution with 280 gal of the 8.4% solution.

69. What is the percent strength of a solution (soln.) mixture if 22 gal of a 3.0% solution are mixed with 56 gal of a 12.3% solution? The density of the both solutions is 8.84 lb/gal.

$$\textbf{Percent mixture strength} = \frac{\textbf{Soln. 1 gal(Available \%/100\%) + Soln. 2 gal(Available \%/100\%)}}{\textbf{Gallons of soln. 1 + Gallons of soln. 2}} \times \textbf{100\% strength}$$

Because both solutions have the same density, these values do not have to be put in the equation as they would just cancel each other out.

$$\text{Percent mixture strength} = \frac{22 \text{ gal}(3.0\%/100\%) + 56 \text{ gal}(12.3\%/100\%)}{22 \text{ gal} + 56 \text{ gal}} \times 100\%$$

$$\text{Percent mixture strength} = \frac{0.66 \text{ gal} + 6.89 \text{ gal}}{78 \text{ gal}} \times 100\% = \frac{755 \text{ gal}}{78 \text{ gal}} = \textbf{9.7\% strength of new solution}$$

70. What percent hypochlorite solution would result if 75 gal of a 24% solution were mixed with 235 gal of a 4.5% solution? Assume both solutions have the same density.

First, find the total volume that would result from mixing these two solutions.

$$\text{Total volume} = 75 \text{ gal} + 235 \text{ gal} = 310 \text{ gal}$$

The equation is:

$$\mathbf{C_1V_1 + C_2V_2 = C_3V_3}$$

where C_1 and C_2 = percent concentration of the two solutions before being mixed
V_1 and V_2 = volume of the two solutions before being mixed
C_3 and V_3 = the resulting percent concentration and volume, respectively

Substitute and divide all concentrations and volumes by 100%.

$$\frac{(24\%)(75 \text{ gal})}{100\%} + \frac{(4.5\%)(235 \text{ gal})}{100\%} = \frac{C_3(310 \text{ gal})}{100\%}$$

Reduce left side of equation by dividing by 100% and then solve for C_3.

$$C_3 = \frac{(18 \text{ gal} + 10.575 \text{ gal})}{310 \text{ gal}}(100\%)$$

$$C_3 = \textbf{9.2\% final solution}$$

71. What percent hypochlorite solution would result if 130 gal of a 9% solution were mixed with 120 gal of a 3% solution? Assume both solutions have the same density.

First, find the total volume that would result from mixing these two solutions.

$$\text{Total volume} = 130 \text{ gal} + 120 \text{ gal} = 250 \text{ gal}$$

The equation is:

$$\mathbf{C_1V_1 + C_2V_2 = C_3V_3}$$

where C_1 and C_2 = percent concentration of the two solutions before being mixed
V_1 and V_2 = volume of the two solutions before being mixed
C_3 and V_3 = the resulting percent concentration and volume, respectively

Substitute and divide all concentrations and volumes by 100%.

$$\frac{(9\%)(130 \text{ gal})}{100\%} + \frac{(3\%)(120 \text{ gal})}{100\%} = \frac{C_3(250 \text{ gal})}{100\%}$$

Reduce left side of equation by dividing by 100% and then solve for C_3.

$$C_3 = \frac{(11.7 \text{ gal} + 3.6 \text{ gal})(100\%)}{250 \text{ gal}}$$

$C_3 = 6.12$, round to **6% final solution**

72. What is the percent strength of a solution (soln.) if 25 gal of a 12% solution is mixed with 75 gal of a 3.0% solution? The 12% solution has a specific gravity of 1.18 and the 3% solution weighs 8.60 lb/gal.

First, determine the weight in pounds of the 12% solution.

$$\text{Weight of 12\% solution} = \text{Specific gravity (8.34 lb/gal)}$$

$$\text{Weight of 12\% solution} = 1.18(8.34 \text{ lb/gal}) = 9.84 \text{ lb/gal}$$

The equation is:

$$\textbf{Percent mixture} =$$

$$\frac{\textbf{Soln. 1 gal(lb/gal)(Avail. \%/100\%) + Soln. 2, gal(lb/gal)(Avail. \%/100\%)}}{\textbf{Soln. 1, gal(lb/gal) + Soln. 2, gal(lb/gal)}} \times \mathbf{100\%}$$

$$= \mathbf{100\%}$$

$$\text{Percent mixture strength} = \frac{25\text{ gal}(9.84\text{ lb/gal})(12\%/100\%) + 75\text{ gal}(8.60\text{ lb/gal})(3.0\%/100\%)}{25\text{ gal}(9.84\text{ lb/gal}) + 75\text{ gal}(8.60\text{ lb/gal})} \times 100\%$$

$$\text{Percent mixture strength} = \frac{(29.52\text{ lb} + 19.35\text{ lb})}{246\text{ lb} + 645\text{ lb}}(100\%) = \frac{(48.87\text{ lb})(100\%)}{891\text{ lb}} = \textbf{5.5\% strength of new solution}$$

73. What percent polymer solution would result if 826 gal of a 25% solution were mixed with 295 gal of a 20% solution?

First, find the total volume that would result from mixing these two solutions.

$$\text{Total volume} = 826\text{ gal} + 295\text{ gal} = 1{,}121\text{ gal}$$

The equation is:

$$\mathbf{C_1V_1 + C_2V_2 = C_3V_3}$$

where C_1 and C_2 = percent concentration of the two solutions before being mixed
V_1 and V_2 = volume of the two solutions before being mixed
C_3 and V_3 = the resulting percent concentration and volume, respectively

Substitute and divide all concentrations and volumes by 100%.

$$\frac{(25\%)(826\text{ gal})}{100\%} + \frac{(20\%)(295\text{ gal})}{100\%} = \frac{C_3(1{,}121\text{ gal})}{100\%}$$

Reduce left side of equation by dividing by 100% and then solve for C_3.

$$C_3 = \frac{(206.5\text{ gal} + 59\text{ gal})(100\%)}{1{,}121\text{ gal}} = \frac{(265.5\text{ gal})(100\%)}{1{,}121\text{ gal}}$$

$C_3 = 23.68$, round to **24% final solution**

74. What is the percent strength of a solution (soln.) if 18 gal of a 21% solution are mixed with 120 gal of a 6.5% solution? The 21% solution has a specific gravity of 1.15 and the 6.5% solution weighs 8.89 lb/gal.

First, determine the weight in pounds of the 21% solution.

$$\text{Weight of 21\% Solution} = \text{Specific gravity (8.34 lb/gal)}$$

$$\text{Weight of 21\% Solution} = 1.15(8.34\ \text{lb/gal}) = 9.59\ \text{lb/gal}$$

The equation is:

$$\textbf{Percent mixture strength} =$$

$$\frac{\textbf{Soln. 1, gal(lb/gal)(Avail. \%/100\%) + Soln. 2, gal(lb/gal)(Avail. \%/100\%)}}{\textbf{Soln. 1, gal(lb/gal) + Soln. 2, gal(lb/gal)}} \times \textbf{100\%}$$

$$\text{Percent mixture strength} =$$

$$\frac{18\ \text{gal}(9.59\ \text{lb/gal})(21\%/100\%) + 120\ \text{gal}(8.89\ \text{lb/gal})(6.5\%/100\%)}{18\ \text{gal}(9.59\ \text{lb/gal}) + 120\ \text{gal}(8.89\ \text{lb/gal})} \times 100\%$$

$$\text{Percent mixture strength} =$$

$$\frac{(36.25\ \text{lb} + 69.342\ \text{lb})(100\%)}{172.62\ \text{lb} + 1{,}066.80\ \text{lb}} = \frac{(105.592\ \text{lb})(100\%)}{1{,}239.42\ \text{lb}}$$

$$= \textbf{8.5\% mixture strength of solution}$$

75. A storage tank contains 260 gal of a 15% zinc orthophosphate (ZOP) solution (soln.) that weighs 10.32 lb/gal. Note that 1,000 gal of a 12.3% ZOP solution are added to the tank. The certificate of analysis on the new ZOP indicates that the specific gravity is 1.12 lb/gal. What is the percent strength of the resulting solution?

First, determine the weight in pounds of the 12.3% solution.

$$\text{Weight of 12.3\% solution} = \text{Specific gravity (8.34 lb/gal)}$$

$$\text{Weight of 12.3\% solution} = 1.12(8.34\ \text{lb/gal}) = 9.34\ \text{lb/gal}$$

The equation is:

$$\textbf{Percent mixture} = \frac{\textbf{Soln. 1, gal(lb/gal)(Avail. \%/100\%) + Soln. 2, gal(lb/gal)(Avail. \%/100\%)}}{\textbf{Soln. 1, gal(lb/gal) + Soln. 2, gal(lb/gal)}} \times \textbf{100\%}$$

$$\text{Percent mixture strength} = \frac{260\text{ gal}(10.32\text{ lb/gal})(15\%/100\%) + 1{,}000\text{ gal}(9.34\text{ lb/gal})(12.3\%/100\%)}{260\text{ gal}(10.32\text{ lb/gal}) + 1{,}000\text{ gal}(9.34\text{ lb/gal})} \times 100\%$$

$$\text{Percent mixture strength} = \frac{(402.48\text{ lb} + 1{,}148.82\text{ lb})(100\%)}{2{,}683.2\text{ lb} + 9{,}340\text{ lb}} = \frac{(1{,}551.30\text{ lb})(100\%)}{12{,}023.2\text{ lb}}$$

$$= \textbf{13\% strength of new ZOP solution}$$

76. What percent polymer solution would result if 375 gal of a 13.6% solution were mixed with 375 gal of a 2.8% solution?

First, find the total volume that would result from mixing these two solutions.

$$\text{Total volume} = 375\text{ gal} + 375\text{ gal} = 750\text{ gal}$$

The equation is:

$$\mathbf{C_1V_1 + C_2V_2 = C_3V_3}$$

where C_1 and C_2 = percent concentration of the two solutions before being mixed
V_1 and V_2 = volume of the two solutions before being mixed
C_3 and V_3 = the resulting percent concentration and volume, respectively

Substitute and divide all concentrations and volumes by 100%.

$$\frac{(13.6\%)(375\text{ gal})}{100\%} + \frac{(2.8\%)(375\text{ gal})}{100\%} = \frac{C_3(750\text{ gal})}{100\%}$$

Solve for C_3.

$$C_3 = \frac{(51\text{ gal} + 10.5\text{ gal})(100\%)}{750\text{ gal}}$$

$$C_3 = \textbf{8.2\% final solution}$$

***Note*: In this case, it is the median value of the two solutions because the volumes mixed are the same.**

77. If 50.0 gal of a 3.00% polymer solution is made, how many pounds of polymer are needed? Assume the polymer weighs 8.34 lb/gal.

The following applies:

$$1\% = 10{,}000 \text{ ppm or mg/L}$$

It follows that:

$$3.00\% = (3)(10{,}000 \text{ mg/L}) = 30{,}000 \text{ mg/L}$$

Note that there is 1 g/1,000 mg, 1 lb in 454 g, and 3.785 L in 1 gal. Arrange the known values so that all the units cancel except the number of pounds.

$$\text{Pounds, polymer} = \frac{(30{,}000 \text{ mg}) \quad (1 \text{ g}) \quad (1 \text{ lb}) \; (3.785 \text{ L}) \; (50.0 \text{ gal})}{(1 \text{ L}) \quad (1{,}000 \text{ mg}) \; (454 \text{ g}) \; (1 \text{ gal})}$$

$$= \mathbf{12.5 \text{ lb of polymer}}$$

78. What is the percent strength of a solution (soln.) if 80 gal of a 17% solution is mixed with 250 gal of a 5% solution? The 17% solution has a specific gravity of 1.07 and the 5% solution weighs 8.85 lb/gal.

First, determine the weight in pounds of the 17% solution.

$$\text{Weight of 17\% solution} = (\text{Specific gravity})(8.34 \text{ lb/gal})$$

$$\text{Weight of 17\% solution} = (1.07)(8.34 \text{ lb/gal}) = 8.92 \text{ lb/gal}$$

The equation is:

$$\textbf{Percent mixture strength} =$$

$$\frac{\textbf{Soln. 1, gal(lb/gal)(Avail. \%/100\%) + Soln. 2, gal(lb/gal)(Avail. \%/100\%)}}{\textbf{Soln. 1, gal(lb/gal) + Soln. 2, gal(lb/gal)}} \times \mathbf{100\%}$$

$$\text{Percent mixture strength} =$$

$$\frac{80 \text{ gal}(8.92 \text{ lb/gal})(17\%/100\%) + 250 \text{ gal}(8.85 \text{ lb/gal})(5\%/100\%)}{80 \text{ gal}(8.92 \text{ lb/gal}) + 250 \text{ gal}(8.85 \text{ lb/gal})} \times 100\%$$

$$\text{Percent mixture strength} = \frac{(121.312 \text{ lb} + 110.625 \text{ lb})(100\%)}{713.6 \text{ lb} + 2{,}212.5 \text{ lb}} = \frac{(231.937 \text{ lb})(100\%)}{2{,}926.1 \text{ lb}}$$

$$= \mathbf{8\% \text{ strength of new solution}}$$

79. What is the approximate percent strength of a hypochlorite solution if 9.6 lb/d of chlorine are required and the hypochlorinator pump delivers 38.6 gpd.

The equation is:

$$\textbf{Hypochlorite solution, \%} = \frac{\textbf{(Chlorine required, pounds per day)(100\%)}}{\textbf{(Hypochlorinator flow, gpd)(8.34 lb/gal)}}$$

Substitute known values and solve.

$$\text{Hypochlorite solution, \%} = \frac{(9.6 \text{ lb/d})(100\%)}{(38.6 \text{ gpd})(8.34 \text{ lb/gal})}$$

Hypochlorite solution, % = 2.98%, round to **3% hypochlorite solution**

ION EXCHANGE SOFTENING PROBLEMS

80. Five ion exchange softeners each have 45.0 ft³ of cationic resin. The capacity of the resin is 32,500 grains/ft³. How many gallons of water will be produced before the softeners are exhausted if the water hardness is 287 mg/L?

First, find the hardness of the water in grains per gallon.

$$\frac{(287 \text{ mg/L})/(17.12 \text{ mg/L})}{1 \text{ gpg}} = 16.764 \text{ gpg}$$

Then, find the exchange capacity.

$$(32{,}500 \text{ grains/ft}^3)(45.0 \text{ ft}^3)(5 \text{ softening units}) = 7{,}312{,}500 \text{ grains}$$

The equation is:

$$\textbf{Water treatment capacity, gal} = \frac{\textbf{Exchange capacity, grains}}{\textbf{Hardness, gpg}}$$

$$\text{Water treatment capacity, gal} = \frac{(7{,}312{,}500 \text{ grains})}{16.764 \text{ gpg}} = 436{,}203 \text{ gal, round to } \textbf{436,000 gal}$$

81. Four ion exchange softeners each have 50.0 ft^3 of cationic resin. The capacity of the resin is 37,800 grains/ft^3. How many gallons of water will be produced before the softeners are exhausted if the water hardness is 62 mg/L?

First, find the hardness of the water in grains per gallon.

$$\frac{(62 \text{ mg/L})/(17.12 \text{ mg/L})}{1 \text{ gpg}} = 3.62 \text{ gpg}$$

Then, find the exchange capacity.

$$(37{,}800 \text{ grains/ft}^3)(50.0 \text{ ft}^3)(4 \text{ softening units}) = 7{,}560{,}000 \text{ grains}$$

The equation is:

$$\textbf{Water treatment capacity, gal} = \frac{\textbf{Exchange capacity, grains}}{\textbf{Hardness, gpg}}$$

$$\text{Water treatment capacity, gal} = \frac{(7{,}560{,}000 \text{ grains})}{3.62 \text{ gpg}}$$

$$= 2{,}088{,}398 \text{ gal, round to } \mathbf{2{,}100{,}000 \text{ gal}}$$

82. An ion exchange softener has 249 ft^3 of cationic resin with a capacity of 24,131 grains/ft^3. If the softener becomes exhausted after softening 594,255 gal, what is the hardness of the water being treated?

First, find the total exchange capacity in grains.

$$\text{Exchange capacity} = (24{,}131 \text{ grains/ft}^3)(249 \text{ ft}^3) = 6{,}008{,}619 \text{ grains}$$

Next, find the hardness in grains per gallon.

The equation is:

$$\textbf{Water treatment capacity, gal} = \frac{\textbf{Exchange capacity, grains}}{\textbf{Hardness, gpg}}$$

Rearrange the equation and solve for hardness in grains per gallon.

$$\text{Hardness, gpg} = \frac{6{,}008{,}619 \text{ grains}}{594{,}255 \text{ gal}} = 10.11 \text{ gpg}$$

Last, calculate the water hardness in milligrams per liter.

$$\text{Water hardness, mg/L} = \frac{(10.11 \text{ mg/L})/(17.12 \text{ mg/L})}{1 \text{ gpg}} = \mathbf{173\ mg/L}$$

83. An ion exchange softener has 48.0 ft^3 of cationic resin with a capacity of 29,500 grains/ft^3. If the softener becomes exhausted after softening 97,500 gal, what is the hardness of the water being treated?

First, find the total exchange capacity in grains.

$$\text{Exchange capacity} = (29{,}500 \text{ grains/ft}^3)(48.0 \text{ ft}^3) = 1{,}416{,}000 \text{ grains}$$

Next, find the hardness in grains per gallon.

The equation is:

$$\textbf{Water treatment capacity, gal} = \frac{\textbf{Exchange capacity, grains}}{\textbf{Hardness, gpg}}$$

Rearrange the equation and solve for hardness in grains per gallon.

$$\text{Hardness, gpg} = \frac{1{,}416{,}000 \text{ grains}}{97{,}500 \text{ gal}} = 14.523 \text{ gpg}$$

Last, calculate the water hardness in milligrams per liter.

$$\text{Water hardness, mg/L} = \frac{(14.523 \text{ mg/L})/(17.12 \text{ mg/L})}{1 \text{ gpg}} = \mathbf{249\ mg/L}$$

84. Determine the ion exchange softener operating time in hours given the following:

Exchange capacity of the softener is 6,575,000.

Flow rate is 180 gpm.

Raw water contains 15.7 gpg.

First, find the capacity of the softener in gallons.

The equation is:

$$\textbf{Water treatment capacity, gal} = \frac{\textbf{Exchange capacity, grains}}{\textbf{Hardness, gpg}}$$

$$\text{Water treatment capacity, gal} = \frac{6{,}575{,}000 \text{ grains}}{15.7 \text{ gpg}} = 418{,}790 \text{ gal}$$

Next, convert flow rate of gallons per minute to gallons per hour.

$$(180 \text{ gpm})(60 \text{ min/hr}) = 10{,}800 \text{ gph}$$

Then, determine the operating time.

$$\textbf{Operating time, hr} = \frac{\textbf{Water treated, gal}}{\textbf{Flow rate, gph}}$$

$$\text{Operating time, hr} = \frac{418{,}790 \text{ gal}}{10{,}800 \text{ gph}} = 38.78 \text{ hr, round to } \mathbf{39\ hr}$$

85. Determine the ion exchange softener operating time in hours given the following:

Exchange capacity of the softener is 8,262,000.

Flow rate is 140 gpm.

Raw water contains 10.1 gpg.

First, find the capacity of the softener in gallons.

The equation is:

$$\textbf{Water treatment capacity, gal} = \frac{\textbf{Exchange capacity, grains}}{\textbf{Hardness, gpg}}$$

$$\text{Water treatment capacity, gal} = \frac{8{,}262{,}000 \text{ grains}}{10.1 \text{ gpg}} = 818{,}020 \text{ gal}$$

Next, convert flow rate of gallons per minute to gallons per hour.

$$(140 \text{ gpm})(60 \text{ min/hr}) = 8{,}400 \text{ gph}$$

Then, determine the operating time.

$$\textbf{Operating time, hr} = \frac{\textbf{Water treated, gal}}{\textbf{Flow rate, gph}}$$

$$\text{Operating time, hr} = \frac{818{,}020 \text{ gal}}{8{,}400 \text{ gph}} = \mathbf{97 \text{ hr}}$$

86. Determine the ion exchange softener operating time in hours given the following:
Exchange capacity of the softener is 6,080,000.
Flow rate is 150 gpm.
Raw water contains 14.2 gpg.

First, find the capacity of the softener in gallons.

The equation is:

$$\textbf{Water treatment capacity, gal} = \frac{\textbf{Exchange capacity, grains}}{\textbf{Hardness, gpg}}$$

$$\text{Water treatment capacity, gal} = \frac{6{,}080{,}000 \text{ grains}}{14.2 \text{ gpg}} = 428{,}169 \text{ gal}$$

Next, convert flow rate of gallons per minute to gallons per hour.

$$(150 \text{ gpm})(60 \text{ min/hr}) = 9{,}000 \text{ gph}$$

Then, determine the operating time.

$$\textbf{Operating time, hr} = \frac{\textbf{Water treated, gal}}{\textbf{Flow rate, gph}}$$

$$\text{Operating time, hr} = \frac{428{,}169 \text{ gal}}{9{,}000 \text{ gph}} = \mathbf{48 \text{ hr}}$$

87. Determine the ion exchange softener operating time in hours given the following:
Exchange capacity of the softener is 4,890,000.
Flow rate is 200 gpm.
Raw water contains 13.8 gpg.

First, find the capacity of the softener in gallons.

The equation is:

$$\textbf{Water treatment capacity, gal} = \frac{\textbf{Exchange capacity, grains}}{\textbf{Hardness, gpg}}$$

$$\text{Water treatment capacity, gal} = \frac{4{,}890{,}000 \text{ grains}}{13.8 \text{ gpg}} = 354{,}348 \text{ gal}$$

Next, convert flow rate of gallons per minute to gallons per hour.

$$(200 \text{ gpm})(60 \text{ min/hr}) = 12{,}000 \text{ gph}$$

Then, determine the operating time.

$$\textbf{Operating time, hr} = \frac{\textbf{Water treated, gal}}{\textbf{Flow rate, gph}}$$

$$\text{Operating time, hr} = \frac{354{,}348 \text{ gal}}{12{,}000 \text{ gph}} = \textbf{30 hr}$$

REVERSE OSMOSIS RECOVERY PERCENTAGE

88. A reverse osmosis unit is being fed at 2,550 gpm. What is the percent recovery if the product flow is 2,145 gpm?

The equation is:

$$\textbf{Percent recovery} = \frac{\textbf{(Product flow)(100\%)}}{\textbf{Feed flow}}$$

$$\text{Percent recovery} = \frac{(2{,}145 \text{ gpm})(100\%)}{2{,}550 \text{ gpm}}$$

$$\text{Percent recovery} = \textbf{84\%}$$

89. Given the following data, what is the percent recovery for a reverse osmosis unit with a 4-2-1 arrangement?

Product flow is 1,186 gpm.

Feed flow is 2.09 mgd.

First, convert million gallons per day to gallons per minute.

$$\textbf{Number of gallons per minute} = \textbf{(mgd)(1,000,000/M)(1 d/1440 min)}$$

$$\text{Number of gallons per minute} = (2.09 \text{ mgd})(1{,}000{,}000/\text{M})(1 \text{ d}/1440 \text{ min})$$
$$= 1{,}451 \text{ gpm}$$

Now, calculate the percent recovery.

The equation is:

$$\textbf{Percent recovery} = \frac{\textbf{(Product flow)(100\%)}}{\textbf{Feed flow}}$$

$$\text{Percent recovery} = \frac{(1{,}186 \text{ gpm})(100\%)}{1{,}451 \text{ gpm}}$$

$$\text{Percent recovery} = \mathbf{81.7\%}$$

PERCENT OF ELEMENT IN A COMPOUND

Atomic weights in this section can be found in Appendix C (Table C-1).

90. What is the percent fluoride (F) ion in fluorosilicic acid (H_2SiF_6)?

The equation for calculating the percent F in H_2SiF_6 is:

$$\textbf{Percent F} = \frac{\textbf{(Molecular weight of F)(100\%)}}{\textbf{Molecular weight of } \mathbf{H_2SiF_6}}$$

First, determine the molecular weight of each of the elements in the compound:

Element	Number of atoms		Atomic weight, g		Molecular weight, g
H	2	×	1.008	=	2.016
Si	1	×	28.06	=	28.06
F	6	×	19.00	=	114.00
			Molecular weight of H_2SiF_6	=	144.076

The molecular weight of F in H_2SiF_6 is 114.00 g.

Substituting:

$$\text{Percent F} = \frac{(114.00 \text{ g})(100\%)}{144.076 \text{ g}} = \mathbf{79.12\% \ F}$$

91. What is the percent copper (Cu) in copper sulfate ($CuSO_4$)?

The equation for calculating the percent Cu in $CuSO_4$ is:

$$\textbf{Percent Cu} = \frac{\textbf{(Molecular weight of Cu)(100\%)}}{\textbf{Molecular weight of } \mathbf{CuSO_4}}$$

First, determine the molecular weight of each of the elements in the compound:

Element	Number of atoms		Atomic weight, g		Molecular weight, g
Cu	1	×	63.54	=	63.54
S	1	×	32.064	=	32.064
O	4	×	15.9994	=	63.9976
			Molecular weight of $CuSO_4$	=	159.6016

The molecular weight of Cu in $CuSO_4$ is 63.54 g.

Substituting:

$$\text{Percent Cu} = \frac{(63.54\text{ g})(100\%)}{159.6016\text{ g}} = \mathbf{39.81\%\ Cu}$$

92. What is the percent Al in Alum $Al_2(SO_4)_3 \cdot 14(H_2O)$? Round atomic weights to nearest 0.01.

The equation for calculating the percent Al in alum [$Al_2(SO_4)_3 \cdot 14(H_2O)$] is:

$$\textbf{Percent Al} = \frac{\textbf{(Molecular weight of Al)(100\%)}}{\textbf{Molecular weight of } \mathbf{Al_2(SO_4)_3}}$$

First, determine the molecular weight of each of the elements in the compound:

Element	Number of atoms		Atomic weight, g		Molecular weight, g
Al	2	×	26.98	=	53.96
S	3	×	32.06	=	96.18
O	26	×	16.00	=	416.00
H	28	×	1.01	=	28.28
	Molecular weight of $Al_2(SO_4)_3 \cdot 14(H_2O)$			=	594.42

The molecular weight of Al in $Al_2(SO_4)_3 \cdot 14(H_2O)$ is 53.96 g.

Substituting:

$$\text{Percent Al} = \frac{(53.96\text{ g})(100\%)}{594.42\text{ g}} = 9.077\%, \text{ round to } \mathbf{9.08\%\ Al}$$

Normally, chemists do not take the water into account.

93. What is the percent manganese (Mn) in potassium permanganate ($KMnO_4$)?

The equation for calculating the percent Mn in $KMnO_4$ is:

$$\textbf{Percent Mn} = \frac{\textbf{Molecular weight of Mn(100\%)}}{\textbf{Molecular weight of } \mathbf{KMnO_4}}$$

First, determine the molecular weight of each of the elements in the compound:

Element	Number of atoms		Atomic weight, g		Molecular weight, g
K	1	×	39.102	=	39.102
Mn	1	×	54.938	=	54.938
O	4	×	15.9994	=	63.9976
			Molecular weight of $CuSO_4$	=	158.0376

The molecular weight of Mn in $KMnO_4$ is 54.938 g.

Substituting:

$$\text{Percent Mn} = \frac{(54.938 \text{ g})(100\%)}{158.0376 \text{ g}} = \mathbf{34.763\% \text{ Mn}}$$

94. What is the percent calcium (Ca) in calcium sulfate ($CaSO_4$)?

The equation for calculating the percent Ca in $CaSO_4$ is:

$$\textbf{Percent Ca} = \frac{\textbf{Molecular weight of Ca(100\%)}}{\textbf{Molecular weight of } \mathbf{CaSO_4}}$$

First, determine the molecular weight of each of the elements in the compound:

Element	Number of atoms		Atomic weight, g		Molecular weight, g
Ca	1	×	40.08	=	40.08
S	1	×	32.064	=	32.064
O	4	×	15.9994	=	63.9976
			Molecular weight of $CuSO_4$	=	136.1416

The molecular weight of Ca in $CaSO_4$ is 40.08 g.

Substituting:

$$\text{Percent Ca} = \frac{(40.08 \text{ g})(100\%)}{136.1416 \text{ g}} = \mathbf{29.44\% \text{ Ca}}$$

CORROSION PROBLEMS

Tables for Corrosion Problems

TABLE 4-1 Temperature with values for A

Temp, C°	A values
0	2.34
5	2.27
10	2.20
15	2.12
20	2.05
25	1.98
30	1.91
35	1.83
40	1.76
45	1.69
50	1.62
55	1.54
60	1.47
65	1.40
70	1.33
75	1.25
80	1.18
85	1.1
90	1.04
95	0.96
100	0.88

TABLE 4-2 Total dissolved solids (TDS) with values for B

TDS, mg/L	B values
0	9.63
50	9.72
100	9.75
200	9.80
400	9.86
800	9.94
1,600	10.04

TABLE 4-3 Calcium or alkalinity as $CaCO_3$, mg/L with log values

Ca or alkalinity as $CaCO_3$ mg/L	Log_{10}
5	
10	1.00
20	1.30
30	1.48
40	1.60
50	1.70
60	1.78
70	1.84
80	1.90
90	1.95
100	2.00
200	2.30
300	2.48
400	2.60
500	2.70
600	2.78
700	2.84
800	2.90
900	2.95
1,000	3.00

Tables 4-1, 4-2, and 4-3 printed by permission of the Office of Water Programs, California State University, Sacramento Foundation. Additional values were extrapolated by the author.

95. Water being tested has the following characteristics: pH is 7.9, temperature is 10°C, TDS is 300, alkalinity is 150, and a calcium hardness of 100. Find the pH_s and the Langelier index (LI). See Tables 4-1, 4-2, and 4-3 for appropriate values.

Write the two equations for this problem:

$$\textbf{Langelier index} = \textbf{pH} - \textbf{pH}_s$$

$$\textbf{pH}_s = \textbf{A} + \textbf{B} - \textbf{log}\ (\textbf{Ca}^{2+}) - \textbf{log(Alkalinity)}$$

First, solve for pH_s using the tables:

$$pH_s = 2.20 + 9.83 - 2.00 - 2.15 = 7.88$$

Note: B (300 TDS) is found by extrapolation. Log (alkalinity) 150 is also found through extrapolation. For example, the B value was extrapolated as follows:

$$\text{Langelier index} = 7.9 - 7.88 = \mathbf{0.02}$$

TDS = 300; this falls between TDS 200 with a value of 9.80 and TDS 400 with a value of 9.86. To extrapolate, 9.86 – 9.80 = 0.06. Then, 0.06/2 = 0.03. Next, add this result to the lowest value: 0.03 + 9.80 = 9.83.

Because the LI is positive, the water is scale forming, but not by very much.

96. Find the pH_s and the LI for water with the following characteristics: pH is 7.6, temperature is 10°C, TDS is 200, alkalinity is 200, and a calcium hardness of 50. See Tables 4-1, 4-2, and 4-3 for appropriate values.

The equations for this problem:

$$\textbf{Langelier index} = \textbf{pH} - \textbf{pH}_s$$

$$\textbf{pH}_s = \textbf{A} + \textbf{B} - \textbf{log}\ (\textbf{Ca}^{2+}) - \textbf{log(Alkalinity)}$$

First, solve for pH_s using the tables:

$$pH_s = 2.20 + 9.80 - 1.70 - 2.30 = 8.00$$

$$\text{Langelier index} = 7.6 - 8.00 = \mathbf{-0.4}$$

Because the LI is negative, the water is corrosive.

97. Find the pH_s and the LI for water with the following characteristics: pH is 7.8, temperature is 5°C, TDS is 400, alkalinity is 200, and a calcium hardness of 60. See Tables 4-1, 4-2, and 4-3 for appropriate values.

The equations for this problem:

$$\textbf{Langelier index} = \textbf{pH} - \textbf{pH}_s$$

$$\textbf{pH}_s = \textbf{A} + \textbf{B} - \textbf{log}\ (\textbf{Ca}^{2+}) - \textbf{log}(\textbf{Alkalinity})$$

First, solve for pH_s using the tables:

$$pH_s = 2.27 + 9.86 - 1.78 - 2.30 = 8.05$$

$$\text{Langelier index} = 7.8 - 8.05 = \mathbf{-0.25}$$

Because the LI is negative, the water is corrosive.

CALCULATIONS FOR CHEMICAL PRECIPITATION SOFTENING PROCESS

Use this table to solve the problems that follow.

TABLE 4-4 Molecular weights of chemical compounds

Compound	Molecular weight
Alkalinity, as $CaCO_3$	100.1
Carbon dioxide, CO_2	44.0
Hardness, as $CaCO_3$	100.1
Hydrated lime, $Ca(OH)_2$	74.1
Magnesium, Mg^{2+}	24.3
Magnesium hydroxide, $Mg(OH)_2$	58.3
Quicklime, CaO	56.1
Soda ash, Na_2CO_3	106.0

98. Determine the hydrated lime, soda ash, and carbon dioxide dose requirements in milligrams per liter for water with the following characteristics:

Characteristic	Source water	Softened water
Total alkalinity, mg/L	225 mg/L as $CaCO_3$	39 mg/L
Total hardness, mg/L	342 mg/L as $CaCO_3$	64 mg/L
CO_2, mg/L	11 mg/L	0 mg/L
Mg^{2+}	31 mg/L	7.0 mg/L
pH	7.0	8.6
Lime purity	94%	

Calculate the hydrated lime required in milligrams per liter. Use an excess lime dosage of 15% (115% or 1.15 in decimal form).

The equation is:

$$\textbf{Hydrated lime feed, mg/L} = \frac{(\boldsymbol{A + B + C + D})(\mathbf{1.15})}{\textbf{Lime purity}}$$

Where A is CO_2 in source water:

$$A = (\text{mg/L as } CO_2)(74.1/44)$$

where 74.1 = molecular weight (MW) of $Ca(OH)_2$.

Substitute known values and solve.

$$A = (11 \text{ mg/L})(74.1/44) = 18.525 \text{ mg/L}$$

Where B is bicarbonate (total) alkalinity removed in softening:

$$B = (\text{mg/L as } CaCO_3 \text{ removed})(74.1/100.1)$$

where 100.1 = MW as $CaCO_3$.

Substitute known values and solve.

$$B = (225 \text{ mg/L} - 39 \text{ mg/L})(74.1/100.1) = 137.69 \text{ mg/L}$$

Where C is hydroxide alkalinity in softened effluent:

$$C = (\text{mg/L as } CaCO_3)(74.1/100.1)$$

In this case, there is no hydroxide alkalinity, therefore, $C = 0$.

Where D is magnesium removed in softening:

$$D = (\text{mg/L as } Mg^{2+})(74.1/24.3)$$

where 24.3 = MW of Mg^{2+}.

$$D = (31 \text{ mg/L} - 7.0 \text{ mg/L})(74.1/24.3) = 73.19 \text{ mg/L}$$

$$\text{Hydrated lime feed, mg/L} =$$

$$\frac{(18.525 \text{ mg/L} + 137.69 \text{ mg/L} + 0 + 73.19 \text{ mg/L})(1.15)}{94\%/100\%} = \mathbf{280\ mg/L\ Ca(OH)_2}$$

Calculate the soda ash required in milligrams per liter. First, find the total hardness removed.

$$\text{Total hardness removed, mg/L as } CaCO_3 =$$

$$\text{Total hardness, mg/L as } CaCO_3 - \text{Total hardness remaining, mg/L as } CaCO_3$$

$$\text{Total hardness removed, mg/L as } CaCO_3 = 342 \text{ mg/L} - 64 \text{ mg/L} = 278 \text{ mg/L}$$

$$\text{Noncarbonate hardness, mg/L as } CaCO_3 =$$

$$\text{Total hardness removed, mg/L as } CaCO_3 - \text{Carbonate hardness, m/L as } CaCO_3$$

$$\text{Noncarbonate hardness, mg/L as } CaCO_3 = 278 \text{ mg/L} - 225 \text{ mg/L} = 53 \text{ mg/L as } CaCO_3$$

$$\text{Soda ash feed, mg/L} = (\text{Noncarbonate hardness, mg/L as } CaCO_3)(106/100.1)$$

where 106 = MW of soda ash.

$$\text{Soda ash feed, mg/L} = (53 \text{ mg/L})(106/100.1) = \mathbf{56\ mg/L\ soda\ ash}$$

Calculate the dosage of CO_2 required for recarbonation.

$$\text{Excess lime, mg/L} = (A + B + C + D)(0.15)$$

$$\text{Excess lime, mg/L} = (18.525 \text{ mg/L} + 137.69 \text{ mg/L} + 0 + 73.19 \text{ mg/L})(0.15) = 34.41 \text{ mg/L}$$

$$\text{Total } CO_2 \text{ feed, mg/L} = (Ca(OH)_2 \text{ excess, mg/L})(44/74.1) + (Mg^{2+} \text{ residual, mg/L})(44/24.3)$$

$$\text{Total } CO_2 \text{ feed, mg/L} = (34.41 \text{ mg/L})(44/74.1) + (7.0 \text{ mg/L})(44/24.3)$$

$$\text{Total } CO_2 \text{ feed, mg/L} = 20.43 \text{ mg/L} + 12.67 \text{ mg/L} = \mathbf{33\ mg/L\ CO_2}$$

99. Determine the hydrated lime, soda ash, and carbon dioxide dose requirements in milligrams per liter for water with the following characteristics:

Characteristic	Source water	Softened water
Total alkalinity, mg/L	213 mg/L as $CaCO_3$	32 mg/L
Total hardness, mg/L	320 mg/L as $CaCO_3$	52 mg/L
CO_2, mg/L	14 mg/L	0 mg/L
Mg^{2+}	24 mg/L	6.0 mg/L
pH	7.2	8.3
Lime purity	92%	

Calculate the hydrated lime required in milligrams per liter. Use an excess lime dosage of 15% (115% or 1.15 in decimal form).

The equation is:

$$\textbf{Hydrated lime feed, mg/L} = \frac{\boldsymbol{(A + B + C + D)(1.15)}}{\textbf{Lime purity}}$$

A = Source water CO_2 expressed as mg/L $CaCO_3$

A = (14 mg/L)(74.1/44) = 23.577 mg/L

B = Bicarbonate (total) alkalinity, mg/L as $CaCO_3$

B = (213 mg/L – 32 mg/L)(74.1/100.1) = 133.987 mg/L

C = Hydroxide (OH) alkalinity (there is none in this example)

Therefore, hydroxide alkalinity, C = 0.

D = Mg removed during softening, as mg/L $CaCO_3$

D = (24 mg/L – 6 mg/L)(74.1/24.3) = 54.889 mg/L

$$\text{Hydrated lime feed, mg/L} = \frac{(23.577 \text{ mg/L} + 133.987 \text{ mg/L} + 0 + 54.889 \text{ mg/L})(1.15)}{92\%/100\%}$$

Hydrated lime feed, mg/L = 265.57 mg/L, round to **270 mg/L, $Ca(OH)_2$**

Calculate the soda ash required in milligrams per liter. First, find the total hardness removed.

Total hardness removed, mg/L as $CaCO_3$ =

Total hardness, mg/L as $CaCO_3$ – Total hardness remaining, mg/L as $CaCO_3$

Total hardness removed, mg/L as $CaCO_3$ = 320 mg/L – 52 mg/L = 268 mg/L

$$\text{Noncarbonate hardness, mg/L as } CaCO_3 =$$
$$\text{Total hardness removed, mg/L as } CaCO_3 - \text{Carbonate hardness, mg/L as } CaCO_3$$
$$\text{Noncarbonate hardness, mg/L as } CaCO_3 = 268 \text{ mg/L} - 213 \text{ mg/L} = 55 \text{ mg/L as } CaCO_3$$

$$\text{Soda ash feed, mg/L} = (\text{Noncarbonate hardness, mg/L as } CaCO_3)(106/100.1)$$

where 106 = MW of soda ash.

$$\text{Soda ash feed, mg/L} = (55 \text{ mg/L})(106/100.1) = \mathbf{58 \text{ mg/L soda ash}}$$

Calculate the dosage of CO_2 required for recarbonation.

$$\text{Excess lime, mg/L} = (A + B + C + D)(0.15)$$
$$\text{Excess lime, mg/L} = (23.577 \text{ mg/L} + 133.987 \text{ mg/L} + 0 + 54.889 \text{ mg/L})(0.15) = 31.868 \text{ mg/L}$$
$$\text{Total } CO_2 \text{ feed, mg/L} = (Ca(OH)_2 \text{ excess, mg/L})(44/74.1) + (Mg^{2+} \text{ residual, mg/L})(44/24.3)$$
$$\text{Total } CO_2 \text{ feed, mg/L} = (31.868 \text{ mg/L})(44/74.1) + (6 \text{ mg/L})(44/24.3)$$
$$\text{Total } CO_2 \text{ feed, mg/L} = 18.923 \text{ mg/L} + 10.864 \text{ mg/L} = \mathbf{30 \text{ mg/L } CO_2}$$

100. Determine the hydrated lime, soda ash, and carbon dioxide dose requirements in milligrams per liter for water with the following characteristics:

Characteristic	Source water	Softened water
Total alkalinity, mg/L	318 mg/L as $CaCO_3$	52 mg/L
Total hardness, mg/L	406 mg/L as $CaCO_3$	70.0 mg/L
CO_2, mg/L	16 mg/L	0 mg/L
Mg^{2+}	23 mg/L	6.0 mg/L
pH	7.1	8.5
Lime purity	93%	

Calculate the hydrated lime required in milligrams per liter. Use an excess lime dosage of 15% (115% or 1.15 in decimal form).

The equation is:

$$\textbf{Hydrated lime feed, mg/L} = \frac{\mathbf{(A + B + C + D)(1.15)}}{\textbf{Lime purity}}$$

(continued)

Where A is CO_2 in source water:

$$A = (\text{mg/L as } CO_2)(74.1/44)$$

where 74.1 = molecular weight (MW) of $Ca(OH)_2$.

Substitute known values and solve.

$$A = (16 \text{ mg/L})(74.1/44) = 26.945 \text{ mg/L}$$

Where B is bicarbonate (total) alkalinity removed in softening:

$$B = (\text{mg/L as } CaCO_3 \text{ removed})(74.1/100.1)$$

where 100.1 = MW as $CaCO_3$.

Substitute known values and solve.

$$B = (318 \text{ mg/L} - 52 \text{ mg/L})(74.1/100.1) = 196.909 \text{ mg/L}$$

Where C is hydroxide alkalinity in softened effluent:

$$C = (\text{mg/L as } CaCO_3)(74.1/100.1)$$

In this case, there is no hydroxide alkalinity, therefore, $C = 0$.

Where D is magnesium removed in softening:

$$D = (\text{mg/L as } Mg^{2+})(74.1/24.3)$$

where 24.3 = MW of Mg^{2+}.

$$D = (23 \text{ mg/L} - 6 \text{ mg/L})(74.1/24.3) = 51.840 \text{ mg/L}$$

$$\text{Hydrated lime feed, mg/L} = \frac{(26.945 \text{ mg/L} + 196.909 \text{ mg/L} + 0 + 51.840 \text{ mg/L})(1.15)}{93\%/100\%}$$

$$\text{Hydrated lime feed, mg/L} = \mathbf{340 \text{ mg/L } Ca(OH)_2}$$

Calculate the soda ash required in milligrams per liter. First, find the total hardness removed.

$$\text{Total hardness removed, mg/L as } CaCO_3 =$$

$$\text{Total hardness, mg/L as } CaCO_3 - \text{Total hardness remaining, mg/L as } CaCO_3$$

$$\text{Total hardness removed, mg/L as } CaCO_3 = 406 \text{ mg/L} - 70 \text{ mg/L} = 336 \text{ mg/L}$$

$$\text{Noncarbonate hardness, mg/L as } CaCO_3 =$$
$$\text{Total hardness removed, mg/L as } CaCO_3 - \text{Carbonate hardness, m/L as } CaCO_3$$
$$\text{Noncarbonate hardness, mg/L as } CaCO_3 = 336 \text{ mg/L} - 318 \text{ mg/L} = 18 \text{ mg/L as } CaCO_3$$

$$\text{Soda ash feed, mg/L} = (\text{Noncarbonate hardness, mg/L as } CaCO_3)(106/100.1)$$

where 106 = MW of soda ash.

$$\text{Soda ash feed, mg/L} = (18 \text{ mg/L})(106/100.1) = \mathbf{19 \text{ mg/L soda ash}}$$

Calculate the dosage of CO_2 required for recarbonation.

$$\text{Excess lime, mg/L} = (A + B + C + D)(0.15)$$

$$\begin{aligned}\text{Excess lime, mg/L} &= (26.945 \text{ mg/L} + 196.909 \text{ mg/L} + 0 + 51.840 \text{ mg/L})(0.15)\\ &= 41.354 \text{ mg/L}\end{aligned}$$

$$\text{Total } CO_2 \text{ feed, mg/L} =$$
$$(Ca(OH)_2 \text{ excess, mg/L})(44/74.1) + (Mg^{2+} \text{ residual, mg/L})(44/24.3)$$
$$\text{Total } CO_2 \text{ feed, mg/L} = (41.354 \text{ mg/L})(44/74.1) + (6.0 \text{ mg/L})(44/24.3)$$
$$\text{Total } CO_2 \text{ feed, mg/L} = 24.556 \text{ mg/L} + 10.864 \text{ mg/L} = \mathbf{35 \text{ mg/L } CO_2}$$

101. Determine the hydrated lime, soda ash, and carbon dioxide dose requirements in milligrams per liter for water with the following characteristics:

Characteristic	Source water	Softened water
Total alkalinity, mg/L	183 mg/L as $CaCO_3$	31 mg/L
Total hardness, mg/L	286 mg/L as $CaCO_3$	58 mg/L
CO_2, mg/L	13 mg/L	0 mg/L
Mg^{2+}	12 mg/L	2.0 mg/L
pH	6.9	8.6
Lime purity	92%	

Calculate the hydrated lime required in milligrams per liter. Use an excess lime dosage of 15% (115% or 1.15 in decimal form).

The equation is:

$$\textbf{Hydrated lime feed, mg/L} = \frac{\mathbf{(A + B + C + D)(1.15)}}{\textbf{Lime purity}}$$

(continued)

A = Source water CO_2 expressed as milligrams per liter $CaCO_3$
A = (13 mg/L)(74.1/44) = 21.893 mg/L
B = Bicarbonate (total) alkalinity, milligrams per liter as $CaCO_3$
B = (183 mg/L – 31 mg/L)(74.1/100.1) = 112.52 mg/L
C = Hydroxide (OH) alkalinity

There is no hydroxide alkalinity, therefore, $C = 0$.

D = Mg removed during softening, as milligrams per liter $CaCO_3$
D = (12 mg/L – 2 mg/L)(74.1/24.3) = 30.494 mg/L

Hydrated lime feed, mg/L =

$$\frac{(21.893 \text{ mg/L} + 112.52 \text{ mg/L} + 0 + 30.494 \text{ mg/L})(1.15)}{92\%/100\%} = \mathbf{206\ mg/L\ Ca(OH)_2}$$

Calculate the soda ash required in milligrams per liter. First, find the total hardness removed.

Total hardness removed, mg/L as $CaCO_3$ =

Total hardness, mg/L as $CaCO_3$ – Total hardness remaining, mg/L as $CaCO_3$

Total hardness removed, mg/L as $CaCO_3$ = 286 mg/L – 58 mg/L = 228 mg/L

Noncarbonate hardness, mg/L as $CaCO_3$ =

Total hardness removed, mg/L as $CaCO_3$ – Carbonate hardness, m/L as $CaCO_3$

Noncarbonate hardness, mg/L as $CaCO_3$ = 228 mg/L – 183 mg/L = 45 mg/L as $CaCO_3$

Soda ash feed, mg/L = (Noncarbonate hardness, mg/L as $CaCO_3$)(106/100.1)

where 106 = MW of soda ash.

Soda ash feed, mg/L = (45 mg/L)(106/100.1) = **48 mg/L soda ash**

Calculate the dosage of CO_2 required for recarbonation.

Excess lime, mg/L = $(A + B + C + D)(0.15)$

Excess lime, mg/L = (21.893 mg/L + 112.52 mg/L + 0 + 30.494 mg/L)(0.15)
= 24.74 mg/L

Total CO_2 feed, mg/L =

($Ca(OH)_2$ excess, mg/L)(44/74.1) + (Mg^{2+} residual, mg/L)(44/24.3)

Total CO_2 feed, mg/L = (24.74 mg/L)(44/74.1) + (2 mg/L)(44/24.3)

Total CO_2 feed, mg/L = 14.69 mg/L + 3.62 mg/L = 18.31 mg/L, round to **18 mg/L CO_2**

CALCULATIONS FOR BICARBONATE, CARBONATE, AND HYDROXIDE ALKALINITY

Use this table to solve the problems that follow.

TABLE 4-5 Alkalinity, mg/L as $CaCO_3$

Results of Titration	Bicarbonate Alkalinity	Carbonate Alkalinity	Hydroxide Alkalinity
P = 0	T	0	0
P is less than ½T	T–2P	2P	0
P = ½T	0	2P	0
P is greater than ½T	0	2T–2P	2P–T
P = T	0	0	0

Where P = phenolphthalein alkalinity and T = total alkalinity.

102. Find the phenolphthalein, total bicarbonate, carbonate, and hydroxide alkalinity for a water sample with the following characteristics based on titration results:

Phenolphthalein (P) alkalinity = 29 mg/L
Total (T) alkalinity = 57 mg/L

First, find the relation between P alkalinity and T alkalinity by dividing the total alkalinity by 2.

$$\frac{57 \text{ mg/L, T alkalinity}}{2} = 28.5$$

Because the P alkalinity is greater than ½ the T alkalinity, the fourth row in the table is used to find the bicarbonate, carbonate, and hydroxide alkalinity.

Bicarbonate alkalinity = **0** from table

Carbonate alkalinity = 2T – 2P

Substitute titration results:

Carbonate alkalinity = 2(57 mg/L) – 2(29 mg/L) = 114 mg/L – 58 mg/L

= **56 mg/L as carbonate alkalinity**

Hydroxide alkalinity = 2P – T

(continued)

Again, substitute titration results:

Hydroxide alkalinity = 2(29 mg/L) – 57 mg/L = 58 mg/L – 57 mg/L

= **1 mg/L as hydroxide alkalinity**

103. Find the phenolphthalein, total bicarbonate, carbonate, and hydroxide alkalinity for a water sample with the following characteristics based on titration results:

Phenolphthalein (P) alkalinity	=	**17 mg/L**
Total alkalinity	=	**47 mg/L**

First, find the relation between P alkalinity and T alkalinity by dividing the total alkalinity by 2.

$$\frac{47 \text{ mg/L, T alkalinity}}{2} = 23.5$$

Because the P alkalinity is less than ½ the T alkalinity, the second row in the table is used.

Bicarbonate alkalinity = T – 2P

Substitute titration results:

Bicarbonate alkalinity = 47 mg/L – 2(17 mg/L) = 47 mg/L – 34 mg/L

= **13 mg/L as bicarbonate alkalinity**

Carbonate alkalinity = 2P

Substitute titration results:

Carbonate alkalinity = 2(17 mg/L) = **34 mg/L as carbonate alkalinity**

Hydroxide alkalinity = **0**

104. Find the phenolphthalein, total bicarbonate, carbonate, and hydroxide alkalinity for a water sample with the following characteristics based on titration results:

Phenolphthalein alkalinity	=	**23 mg/L**
Total alkalinity	=	**71 mg/L**

First, find the relation between P alkalinity and T alkalinity by dividing the total alkalinity by 2.

$$\frac{71 \text{ mg/L, T alkalinity}}{2} = 35.5$$

Because the P alkalinity is less than ½ the T alkalinity, the second row in the alkalinity table is used.

$$\text{Bicarbonate alkalinity} = \text{T} - 2\text{P}$$

Substitute titration results:

$$\text{Bicarbonate alkalinity} = 71 \text{ mg/L} - 2(23 \text{ mg/L}) = 71 \text{ mg/L} - 46 \text{ mg/L}$$

$$= \mathbf{25\ mg/L\ as\ bicarbonate\ alkalinity}$$

$$\text{Carbonate alkalinity} = 2\text{P}$$

Substitute titration results:

$$\text{Carbonate alkalinity} = 2(23 \text{ mg/L}) = \mathbf{46\ mg/L\ as\ carbonate\ alkalinity}$$

$$\text{Hydroxide alkalinity} = \mathbf{0}$$

105. Find the phenolphthalein, total bicarbonate, carbonate, and hydroxide alkalinity for a water sample with the following characteristics based on titration results:

Phenolphthalein (P) alkalinity	=	**20 mg/L**
Total alkalinity	=	**40 mg/L**

First, find the relation between P alkalinity and T alkalinity by dividing the total alkalinity by 2.

$$\frac{40 \text{ mg/L, T alkalinity}}{2} = 20$$

(continued)

Because the P alkalinity equals ½ the T alkalinity, the third row in the table is used.

Bicarbonate alkalinity = **0** from table

Carbonate alkalinity = 2P

Substitute titration results:

Carbonate alkalinity = 2(20 mg/L) = **40 mg/L as carbonate alkalinity**

Hydroxide alkalinity = **0** from table

106. Find the phenolphthalein, total bicarbonate, carbonate, and hydroxide alkalinity for a water sample with the following characteristics based on titration results:

Phenolphthalein (P) alkalinity	=	**31 mg/L**
Total alkalinity	=	**42 mg/L**

First, find the relation between P alkalinity and T alkalinity by dividing the total alkalinity by 2.

$$\frac{42 \text{ mg/L, T alkalinity}}{2} = 21$$

Because the P alkalinity is greater than ½ the T alkalinity, the fourth row in the table is used.

Bicarbonate alkalinity = **0** from table

Carbonate alkalinity = 2T – 2P

Substitute titration results:

Carbonate alkalinity = 2(42 mg/L) – 2(31 mg/L) = **22 mg/L as carbonate alkalinity**

Hydroxide alkalinity = 2P – T

Substitute titration results:

Hydroxide alkalinity = 2(31 mg/L) – 42 mg/L = 62 mg/L – 42 mg/L

= **20 mg/L as hydroxide alkalinity**

LOG REMOVAL AND CT PROBLEMS

107. What is the log removal for a water treatment plant if the samples show a raw water coliform count of 280/100 mL (through extrapolation) and the finished water shows 2.0/100 mL?

First, determine percent removal.

$$\textbf{Percent removal} = \frac{\textbf{(In – Out)(100\%)}}{\textbf{In}}$$

$$\text{Percent removal} = \frac{(280 - 2.0)(100\%)}{280} = 99.3\%.$$

Next, change 99.3% to decimal form by dividing by 100% = 0.993.

Then, calculate the log removal.

$$\textbf{Log removal} = \textbf{(Log}_{10}\textbf{)(–)(100.0 – Percent removed in decimal form)}$$

$$\text{Log removal} = (\text{Log}_{10})(-)(100.0 - 0.993) = (\text{Log}_{10}\ 0.007)(-) = (-2.15)(-)$$

= **2.15 logs removed**

108. What is the log removal for a water treatment plant if the samples show a raw water coliform count of 120/100 mL and the finished water shows 1.5/100 mL?

First, determine percent removal.

$$\textbf{Percent removal} = \frac{\textbf{(In – Out)(100\%)}}{\textbf{In}}$$

$$\text{Percent removal} = \frac{(120 - 1.5)(100\%)}{120} = 98.75\%.$$

Next, change 98.75% to decimal form by dividing by 100% = 0.9875.

Then, calculate the log removal.

$$\textbf{Log removal} = \textbf{(Log}_{10}\textbf{)(–)(100.0 – Percent removed in decimal form)}$$

$$\text{Log removal} = (\text{Log}_{10})(-)(100.0 - 0.9875) =$$

$(\text{Log}_{10}\ 0.0125)(-) = (\ 1.9)(-) = 1.903$ logs, round to **1.9 logs removed**

109. Calculate the CT and inactivation ratio for a water treatment plant given the following parameters. Also, determine if this treatment facility meets the CT.

Daily Parameters:

Detention time = 130 min
pH = 7.4
Temperature = 18°C
Lowest chlorine residual = 0.50 mg/L
A 2.0 log removal is required for this system.

First, determine the CT table value.

Determine the CT table value using the 2.0 log removal table (disk included in this book). Find the chart with a pH of 7.4, go down the left column and find the temperature of 18°C. Then go over to 0.5-mg/L free chlorine residual. Because 0.5 is not shown on the chart, extrapolate between the numbers that intersect the temperature and the chlorine residuals of 0.4 and 0.6. To extrapolate, subtract 35.2 (the 0.6 residual) from 34.2 (the 0.4 residual), divide by 2, and add this result to the lowest chlorine residual CT table value.

That is,

$$\text{CT table value} = \frac{35.2 - 34.2}{2} + 34.2 = 34.7 \text{ CT table value}$$

Next, calculate the CT.

$$\textbf{Calculated CT} = \textbf{(Chlorine concentration, mg/L)(Time, min)}$$

$$\text{Calculated CT} = (0.5 \text{ mg/L})(130 \text{ min}) = 65 \text{ is calculated CT}$$

Next, calculate the inactivation ratio.

$$\textbf{Inactivation ratio} = \frac{\textbf{Calculated CT value}}{\textbf{CT table value}}$$

$$\text{Inactivation ratio} = \frac{65}{34.7} = \mathbf{1.87}$$

Because the inactivation ratio value is greater than 1.0, this system meets the CT criteria and is in compliance.

110. Calculate the CT and inactivation ratio for a water treatment plant that has the following parameters. Also, determine if this treatment facility meets the CT.

Daily Parameters:

Detention time = 72 min
pH = 7.6
Temperature = 14°C
Lowest chlorine residual = 0.40 mg/L
A 1.5 log removal is required for this system.

First, determine the CT table value.

Determine CT table value using the 1.5 log removal table (disk included in this book). Find the chart with a pH of 7.6, go down the left column and find a temperature of 14°C. Then go over to 0.4-mg/L free chlorine residual. The number that intersects the temperature and the chlorine residual is the CT table value. In this case, the CT table value is 48.

Next, calculate the CT.

$$\textbf{Calculated CT} = \textbf{(Chlorine concentration, mg/L)(Time, min)}$$

$$\text{Calculated CT} = (0.4 \text{ mg/L})(72 \text{ min}) = 28.8 \text{ is CT calculated}$$

Next, calculate the inactivation ratio.

$$\textbf{Inactivation ratio} = \frac{\textbf{Calculated CT value}}{\textbf{CT table value}}$$

$$\text{Inactivation ratio} = \frac{28.8}{48} = \mathbf{0.6}$$

Because the inactivation ratio value is less than 1.0, this system does not meet the CT criteria and is not in compliance.

111. A conventional water treatment plant had to discontinue prechlorination, that is, they stopped adding chlorine to the flocculation basins and the sedimentation basin due to elevated trihalomethane levels. Consequently, the chlorine dose was increased before the filters and the clear well and a lithium chloride tracer study was performed. The plant requires a 1.0 log removal for *Giardia* cysts. Given the following parameters on the first day of this process change, determine if this plant is in CT compliance.

UNIT PROCESS OR PIPING	T_{10} VALUE, min	LOWEST CHLORINE RESIDUAL, mg/L
Filtration	**12**	**0.4**
Piping (filter to clear well)	**3**	**0.4**
Clear well	**41**	**1.0**

UNIT PROCESS OR PIPING	TEMPERATURE	pH	CT VALUE, TABLES
Filtration	**14**	**6.7**	**35**
Piping (filter to clear well)	**14**	**6.7**	**35**
Clear well	**17**	**7.8**	**45.5**

First, calculate the CT for each.

$$\text{Filtration CT} = (12\text{ min})(0.4\text{ mg/L}) = 4.8$$

$$\text{Piping CT} = (3\text{ min})(0.4\text{ mg/L}) = 1.2$$

$$\text{Clear well CT} = (41\text{ min})(1.0\text{ mg/L}) = 41$$

Next, calculate the inactivation ratio (IR).

The equation is:

$$\textbf{IR} = \frac{\textbf{Calculated CT}}{\textbf{CT value, table}}$$

$$\text{Filtration IR} = \frac{(4.8\text{ calculated, CT})}{(35\text{, CT value, table})} = 0.137$$

$$\text{Piping IR} = \frac{(1.2\text{ calculated, CT})}{(35\text{, CT value, table})} = 0.034$$

$$\text{Clear well IR} = \frac{(41\text{ calculated, CT})}{(45.5\text{ CT value, table})} = 0.901$$

Last, add the three inactivation ratios.

$$\text{IR, total} = 0.137 + 0.034 + 0.901 = 1.072\text{, round to 1}$$

You can argue that it is greater than 1 before rounding.

Because the inactivation ratio is greater than 1.0, the water plant is in compliance with the required CT criteria.

The CT tables found on the CD included with this book and the two tables presented below can be used to solve the following CT problems.

TABLE 4-6 Filtration log removal credits

	Minimum Log Removal Credits	
Type of Treatment	***Giardia***	**Viruses**
Conventional filtration	2.5	2.0
Direct filtration	2.0	1.0
Diatomaceous earth filtration	2.0	1.0
Slow sand filtration	2.0	1.0

TABLE 4-7 Conversion factors of hydraulic detention time to disinfection contact time

Baffling Condition and Description	Conversion Factor
Perfect (plug flow): Very high length-to-width ratio, e.g., pipeline; perforated inlet, outlet, and intrabasin baffles.	1.0
Superior baffling: Perforated inlet baffle serpentine, or perforated launders, or perforated intrabasin baffles and outlet weir.	0.7
Average baffling: Either inlet or outlet baffling with some intrabasin baffles.	0.5
Poor baffling: One or multiple unbaffled inlets and outlets; no intrabasin baffles.	0.3
No baffling, mixed flow: Very low length-to-width ratio; high inlet and outlet flow velocities.	0.1

112. Calculate the CT for the following direct filtration plant to determine if it meets the Surface Water Treatment Rule (SWTR).

Clear well capacity = 2.4 mil gal

pH = 7.5

Flow = 5.3 mgd

Temperature = 15°C

There is a 14-in. pipeline 2,453 ft to first customer.

Chlorine residual at outlet of clear well is 0.8 mg/L.

Chlorine residual at first customer is 0.4 mg/L and pH is 7.5.

Clear well has no baffling and tracer studies indicate T_{10} = 30 min.

First, the SWTR requires a 3 log removal of *Giardia* cysts. From Table 4-6, a plant with direct filtration has 2 log credits. Thus, for this problem we need at least 1.0 log removal to be in compliance.

Normally you would find the detention time (DT) in minutes for the clear well as shown below.

DT = Volume/Flow

$$\text{DT} = \frac{(2{,}400{,}000)(1{,}440 \text{ min})}{(5{,}300{,}000 \text{ gpd})(\text{day})} = 652 \text{ min (absolute ideal conditions)}$$

However, in this problem, a tracer study was done to determine the DT, and it has a value of 30 min. The above DT was determined to show how important baffling could be in a clear well.

Then, calculate the CT.

$$\text{CT}_{\text{calc}} = (\text{Conversion factor})(\text{DT in min})(\text{Cl}_2 \text{ residual})$$

Also, the conversion factor can be eliminated since the actual detention time was determined by a tracer study.

$$\text{CT}_{\text{calc}} = (30 \text{ min})(0.8 \text{ mg/L}) = 24 \text{ CT}_{\text{calc}}$$

Using the disk included in this book, find the CT_{req}. First, find the 1.0 log inactivation table with a pH of 7.5. Then, find the temperature of 15°C in the first column. Last, go to the column with the chlorine residual for the clear well, 0.8 mg/L. In this case it is 29.

$$\text{Inactivation ratio (IR)} = \text{CT}_{\text{calc}}/\ \text{CT}_{\text{req}} = 24/29 = 0.83 \text{ IR for the clear well}$$

Now, determine the CT for the pipeline.

Find the diameter, in feet, for the 14-in. pipeline.

$$\text{Diameter} = \frac{(14 \text{ in.})(1 \text{ ft})}{12 \text{ in.}} = 1.167 \text{ ft}$$

Next, determine the pipe volume.

$$\mathbf{Volume = (0.785)(Diameter)^2(Length)}$$

$$\text{Volume} = (0.785)(1.167 \text{ ft})(1.167 \text{ ft})(2{,}453 \text{ ft}) = 2{,}622.46 \text{ ft}^3$$

Then, determine the number of gallons.

$$\text{Number of gallons} = (2{,}622.46 \text{ ft}^3)(7.48 \text{ gal/ft}^3) = 19{,}616 \text{ gal}$$

Then, determine the detention time (DT).

$$\text{DT} = \text{volume/flow} = \frac{19{,}616 \text{ gal } (1{,}440 \text{ min/d})}{5{,}300{,}000 \text{ gpd}} = 5.33 \text{ min}$$

Last, solve for CT_{calc}.

$$CT_{calc} = (1.0)(5.33 \text{ min})(0.4 \text{ mg/L}) = 2.13\ CT_{calc}$$

Using the disk provided, find the CT_{req}. Go to log inactivation table of 1.0 with a pH of 7.5, temperature of 15°C, and a chlorine residual of 0.4 mg/L. In this case, it is 28.

Inactivation ratio (IR) = $\mathbf{CT_{calc}/\ CT_{req}}$ = 2.13/28 = 0.076 IR for the pipeline

Now, add the two inactivation ratios (clear well and pipeline).

0.83 + 0.076 = 0.906 IR, round to **0.9 IR**

This plant does not meet the 1.0 IR required.

113. Calculate the CT for a slow sand filtration plant with the following characteristics to determine if it meets the Surface Water Treatment Rule (SWTR).

Clear well capacity = 10 mil gal with poor baffling

pH = 7.5 for clear well and distribution system

Chlorine residual at outlet of clear well is 1.5 mg/L.

Temperature = 10°C

12-in. pipeline 875 ft to first customer

Flow = 2.4 mgd

Chlorine residual at first customer is 1.0 mg/L.

Clear well has no baffling and tracer studies indicate T_{10} = 255 min.

First, the SWTR requires a 3 log removal of *Giardia* cysts. From Table 4-6, a plant with slow sand filtration has 2 log credits. Thus, we need at least 1.0 log removal to be in compliance.

Then, calculate the CT.

$$CT_{calc} = (\text{Conversion factor})(\text{DT, in min})(Cl_2 \text{ residual})$$

The tracer study gives the detention time, and it has a value of 255 min (this eliminates the conversion factor).

$$CT_{calc} = (255 \text{ min})(1.5 \text{ mg/L}) = 382.5 \ CT_{calc}$$

Using the disk included in this book, find the CT_{req}. Find the log inactivation table of 1.0 with a pH of 7.5. Now, go to 10°C and the chlorine residual of 1.5 mg/L. Because 1.4 mg/L = 47 and 1.6 mg/L = 48, it follows that 1.5 mg/L falls in the middle of these two numbers.

$$CT_{req} = (47 + 48)/2 = 47.5$$

$$\textbf{Inactivation ratio (IR)} = \mathbf{CT_{calc}/\ CT_{req}}$$

$$= 382.5/47.5 = 8.05 \text{ IR for the clear well}$$

Now, determine the CT for the pipeline.

For a change, use the radius instead of the diameter for the pipeline.

$$\text{Radius} = D/2 = \frac{(12 \text{ in.})\ (1 \text{ ft})}{2 \quad 12 \text{ in.}} = 0.5 \text{ ft}$$

The equation is:

$$\textbf{Volume} = \boldsymbol{\pi r^2}\textbf{(Length)}$$

$$\text{Volume} = (3.14)(0.5 \text{ ft})(0.5 \text{ ft})(875 \text{ ft}) = 687 \text{ ft}^3$$

$$\text{Number of gallons} = (687 \text{ ft}^3)(7.48 \text{ gal/ft}^3) = 5{,}139 \text{ gal}$$

$$DT = \text{volume/flow} = (5{,}139 \text{ gal})(1{,}440 \text{ min/d})/2{,}400{,}000 \text{ gpd} = 3.1 \text{ min}$$

$$CT_{calc} = (1.0 \text{ conversion factor})(3.1 \text{ min})(1.0 \text{ mg/L}) = 3.1 \ CT_{calc}$$

Using the disk provided, find the CT_{req}. Find the log inactivation table of 1.0 with a pH of 7.5. Go to 10°C and the chlorine residual of 1.0 mg/L. In this case it is 45.

$$IR = CT_{calc}/\ CT_{req} = 3.1/45 = 0.07 \text{ IR for the pipeline}$$

Now, add the two inactivation ratios (clear well and pipeline).

$$8.05 + 0.07 = 8.12 \text{ IR, round to } \mathbf{8.1\ IR}$$

This meets the 1.0 IR required.

114. Calculate the CT for a direct filtration plant with the following characteristics to determine if it meets the Surface Water Treatment Rule (SWTR).

Read this problem carefully. There is some information that you can eliminate.

pH = 7.8

Flow = 4.4 mgd

Temperature = 16°C

Clear well capacity = 8 mil gal

Chlorine residual at outlet of clear well is 1.3 mg/L.

Clear well has no baffling and tracer studies indicate T_{10} = 165 min.

Water goes down two 12-in. pipes at a flow of 3.2 mgd and each feeds different covered reservoirs that are each about 1 mile away. From the clear well a 12-in. pipe 345 ft long feeds a storage tank with 750,000 gal. The pH is 7.5 and the temperature is 16°C.

pH = 7.5

Chlorine residual at tank inlet is 1.1 mg/L.

Chlorine residual at tank outlet is 0.8 mg/L.

Pipeline from tank to first customer is 624 ft long and 12 in. in diameter.

Chlorine residual at first customer is 0.6 mg/L from water originating from the storage tank.

First, the SWTR requires a 3 log removal of *Giardia* cysts. From Table 4-6, a plant with direct filtration has 2 log credits. Thus, we need at least 1.0 log removal to be in compliance.

Then, calculate the CT.

$$CT_{calc} = \text{(Conversion factor)(DT, in min)}(Cl_2 \text{ residual})$$

In this problem, a tracer study was done to determine the DT, and it has a value of 165 min. The conversion factor can thus be dropped.

$$CT_{calc} = (165 \text{ min})(1.3 \text{ mg/L}) = 214.5\ CT_{calc}$$

(continued)

Using the disk provided, find the CT_{req}. Find the log inactivation table of 1.0 with a pH of 7.8. Go to 16°C and the chlorine residual of 1.3 mg/L. There is no 1.3 mg/L. However, 1.3 falls between 1.2 mg/L and 1.4 mg/L. Because 1.2 mg/L = 32.9 and 1.4 mg/L = 33.4, it follows that 1.3 mg/L falls in the middle of these two numbers.

$$CT_{req} = (33.4 + 32.9)/2 = 33.15$$

$$\textbf{Inactivation ratio} = \mathbf{CT_{calc}/CT_{req}}$$

$$= 214.5/33.15 = 6.47 \text{ IR for the clear well}$$

Now, determine the CT for the pipeline from the clear well to the tank.

Find the diameter, in feet, for the pipeline.

$$\text{Diameter} = \frac{(12 \text{ in.})(1 \text{ ft})}{12 \text{ in.}} = 1.0 \text{ ft}$$

The equation for volume is:

$$\textbf{Volume} = \mathbf{(0.785)(Diameter)^2(Length)}$$

$$\text{Volume} = (0.785)(1.0 \text{ ft})(1.0 \text{ ft})(345 \text{ ft}) = 270.8 \text{ ft}^3$$

$$\text{Number of gallons} = (270.8 \text{ ft}^3)(7.48 \text{ gal/ft}^3) = 2{,}025.6 \text{ gal}$$

Total flow is 4.4 mgd, but 3.2 mgd is going to the covered reservoirs. Therefore, 1.2 mgd is going to the tank.

$$DT = \text{Volume/Flow} = (2{,}025.6 \text{ gal})(1{,}440 \text{ min/d})/1{,}200{,}000 \text{ gpd} = 2.43 \text{ min}$$

$$CT_{calc} = (1.0 \text{ conversion factor})(2.43 \text{ min})(1.1 \text{ mg/L}) = 2.673\ CT_{calc}$$

From the disk provided, find the CT_{req}. Find the log inactivation table of 1.0 with a pH of 7.5. Go to 16°C and the chlorine residual of 1.1 mg/L (residual just before the tank). The CT required falls between 28.4 and 29.4.

$$CT_{req} = (29.4 + 28.4)/2 = 28.9$$

$$IR = CT_{calc}/\ CT_{req} = 2.673/28.9 = 0.09 \text{ IR for the pipeline}$$

Now, calculate the CT for the tank.

$$DT = \text{volume/flow} = (750{,}000 \text{ gal})(1{,}440 \text{ min/d})/1{,}200{,}000 \text{ gpd} = 900 \text{ min}$$

$$CT_{calc} = (0.1 \text{ conversion factor})(900 \text{ min})(0.8 \text{ mg/L}) = 72\ CT_{calc}$$

From the disk provided, find the CT_{req}. Find the log inactivation table of 1.0 with a pH of 7.5. Now, go to 16°C and the chlorine residual of 0.8 mg/L. In this case, it is 27.6.

$$IR = CT_{calc}/CT_{req} = 72/27.6 = 2.6 \text{ IR for the tank}$$

Find the volume of the pipeline to the first customer.

$$\text{Volume} = \pi r^2(\text{Length})$$

Next, find the radius.

$$\text{Radius} = D/2 = \frac{(12 \text{ in.})\ (1 \text{ ft})}{2 \quad 12 \text{ in.}} = 0.5 \text{ ft}$$

$$\text{Volume} = (3.14)(0.5\ \text{ft})(0.5\ \text{ft})(624\ \text{ft}) = 490\ \text{ft}^3$$

$$\text{Number of gallons} = (490\ \text{ft}^3)(7.48\ \text{gal/ft}^3) = 3{,}665\ \text{gal}$$

$$\text{DT} = \text{volume/flow} = (3{,}665\ \text{gal})(1{,}440\ \text{min/d})/1{,}200{,}000\ \text{gpd} = 4.4\ \text{min}$$

$$\text{CT}_{\text{calc}} = (1.0\ \text{conversion factor})(4.4\ \text{min})(0.6\ \text{mg/L}) = 2.64\ \text{CT}_{\text{calc}}$$

From the disk provided, find the CT_{req}. Find the log inactivation table of 1.0 with a pH of 7.5. Go to 16°C and the chlorine residual of 0.6 mg/L. In this case, it is 27.4.

$$\text{IR } 5\ \text{CT}_{\text{calc}}/\ \text{CT}_{\text{req}} = 2.64/27.4\ 5\ 0.10\ \text{IR for the second pipeline}$$

Now, add the inactivation ratios for each part.

$$6.47 + 0.09 + 2.6 + 0.10 = 9.26\ \text{IR, round to } \mathbf{9.3\ IR}$$

This meets the CT required.

115. Calculate the CTs for a conventional filter plant with the following characteristics to determine if it meets the Surface Water Treatment Rule (SWTR).

Summer parameters:

pH = 7.8

Flow = 15.5 mgd

Temperature = 14°C

Clear well capacity = 10 mil gal with no baffling

Chlorine residual at outlet of clear well is 0.6 mg/L.

One 2-ft diameter pipeline and one 3-ft diameter pipeline feed the distribution system from the clear well. The 2-ft diameter pipeline has already been shown to meet the SWTR for CTs.

Water flows down the 3-ft diameter pipeline at a maximum of 4,960 gpm and feeds into a 1-ft diameter pipeline 248 ft from the clear well. Water flows down the 1-ft diameter pipe at a maximum of 225 gpm and feeds the first customer 67 ft from the 3-ft diameter pipeline.

The chlorine residual for the first customer is 0.3 mg/L.

Winter parameters:

pH = 7.6

Flow = 6.2 mgd

Temperature = 7°C

Chlorine residual from the clear well outlet structure is 0.9 mg/L.

Water flows down the 3-ft diameter pipeline at a maximum of 2,850 gpm and the 1-ft diameter pipeline at a maximum of 110 gpm.

The chlorine residual for the first customer is 0.5 mg/L. Assume the chlorine residual is the same at the end of the 3-ft pipeline.

(continued)

Two separate CTs need to be calculated, one for the summer and the other for the winter. The SWTR requires a 3 log removal of *Giardia* cysts. From Table 4-6, a plant with conventional filtration has 2.5 log credits. Thus, we need at least 0.5 log removal to be in compliance.

First, find the capacity, in gallons, for the clear well and pipelines to the first customer.

$$\text{Clear well, in gal} = (10 \text{ mil gal})(1{,}000{,}000/1 \text{ M}) = 10{,}000{,}000 \text{ gal}$$

The 3-ft diameter pipeline to branch off to the 1-ft diameter pipeline:

The equation is:

$$\textbf{Volume} = \mathbf{(0.785)(Diameter)^2(Length)(7.48\ gal/ft^3)}$$

$$\text{Volume of 3 ft, in gal} = (0.785)(3 \text{ ft})(3 \text{ ft})(248 \text{ ft})(7.48 \text{ gal/ft}^3) = 13{,}106 \text{ gal}$$

1-ft diameter pipeline from 3 ft to first customer:

$$\text{Volume of 1 ft, in gal} = (0.785)(1 \text{ ft})(1 \text{ ft})(67 \text{ ft})(7.48 \text{ gal/ft}^3) = 393 \text{ gal}$$

Next, convert the flow rate from the clear well to gallons per minute.

$$\text{Summer flow rate} = (15.5 \text{ mgd})(1{,}000{,}000/1 \text{ M})(1 \text{ d}/1{,}440 \text{ min}) = 10{,}764 \text{ gpm}$$

$$\text{Winter flow rate} = (6.2 \text{ mgd})(1{,}000{,}000/1 \text{ M})(1 \text{ d}/1{,}440 \text{ min}) = 4{,}306 \text{ gpm}$$

Flow rates for both pipelines for summer and winter are given.

Calculate CT for the clear well.

$$\begin{aligned}\text{Summer } CT_{calc} &= (0.1 \text{ conversion factor})(0.6 \text{ mg/L})(10{,}000{,}000 \text{ gal}/10{,}764 \text{ gpm}) \\ &= 55.7\ CT_{calc}\end{aligned}$$

From the disk provided, find the CT_{req}. Find the log inactivation table of 0.5 with a pH of 7.8. Go to 14°C and the chlorine residual of 0.6 mg/L. In this case, it is 17.4.

$$\textbf{Inactivation ratio (IR)} = \mathbf{CT_{calc}/\ CT_{req}}$$

$$= 55.7/17.4 = 3.2 \text{ IR for the clear well}$$

$$\begin{aligned}\text{Winter } CT_{calc} &= (0.1 \text{ conversion factor})(0.9 \text{ mg/L})(6{,}200{,}000 \text{ gal}/4{,}306 \text{ gpm}) \\ &= 130\ CT_{calc}\end{aligned}$$

From the disk provided, find the CT_{req}. Find the log inactivation table of 0.5 with a pH of 7.6. Go to 7°C and the chlorine residual of 0.9 mg/L. Because 0.9 mg/L is between 0.8 mg/L and 1.0 mg/L, extrapolate between these two values.

$$0.8 \text{ mg/L} = 27.2 \text{ and } 1.0 \text{ mg/L} = 27.9$$

To extrapolate, add these two numbers to each other and divide by 2.

$$(27.9 + 27.2)/2 = 27.55$$

$$IR = CT_{calc}/CT_{req} = 130/27.55 = 4.72 \text{ IR for the clear well}$$

Calculate CT for the 3-ft pipeline.

$$\text{Summer } CT_{calc} = (1.0 \text{ conversion factor})(0.3 \text{ mg/L})(13{,}106 \text{ gal}/4{,}960 \text{ gpm}) = 0.79\ CT_{calc}$$

From the disk provided, find the CT_{req}. Find the log inactivation table of 0.5 with a pH of 7.8. Go to 14°C and the chlorine residual of 0.3 mg/L. Because 0.3 mg/L is less than the chart value, use the value for 0.4 mg/L, which is 17.3.

$$IR = CT_{calc}/\ CT_{req} = 0.79/17.3 = 0.04 \text{ IR for the 3-ft pipeline}$$

$$\text{Winter } CT_{calc} = (1.0 \text{ conversion factor})(0.5 \text{ mg/L})(13{,}106 \text{ gal}/2{,}850 \text{ gpm}) = 2.30\ CT_{calc}$$

From the disk provided, find the CT_{req}. Find the log inactivation table of 0.5 with a pH of 7.6. Go to 7°C and the chlorine residual of 0.5 mg/L. Because 0.5 mg/L is between 0.4 mg/L and 0.6 mg/L, extrapolate between these two values.

$$0.4 \text{ mg/L} = 26.1 \text{ and } 0.6 \text{ mg/L} = 26.8$$

To extrapolate, add these two numbers to each other and divide by 2.

$$CT_{req} = (26.8 + 26.1)/2 = 26.45$$

$$IR = CT_{calc}/CT_{req} = 2.30/26.45 = 0.09 \text{ IR for the 3-ft pipeline}$$

Calculate CT for the 1-ft pipeline.

$$\text{Summer } CT_{calc} = (1.0 \text{ conversion factor})(0.3 \text{ mg/L})(393 \text{ gal}/225 \text{ gpm}) = 0.52\ CT_{calc}$$

From the disk provided, find the CT_{req}. Find the log inactivation table of 0.5 with a pH of 7.8. Go to 14°C and the chlorine residual of 0.3 mg/L. Note that 0.3 mg/L is less than the chart value, so use the 0.4-mg/L value of 17.3.

$$IR = CT_{calc}/\ CT_{req} = 0.52/17.3 = 0.03 \text{ IR for the 1-ft pipeline}$$

$$\text{Winter } CT_{calc} = (1.0 \text{ conversion factor})(0.5 \text{ mg/L})(393 \text{ gal}/110 \text{ gpm}) = 1.79\ CT_{calc}$$

From the disk provided, find the CT_{req}. Find the log inactivation table of 0.5 with a pH of 7.6. Go to 7°C and the chlorine residual of 0.5 mg/L. Because 0.5 mg/L lies between 0.4 mg/L and 0.6 mg/L, extrapolate. 0.4 mg/L = 26.1 and 0.6 mg/L = 26.8. To extrapolate, add these two numbers to each other and divide by 2.

$$(26.8 + 26.1)/2 = 26.45$$

$$IR = CT_{calc}/CT_{req} = 1.79/26.45 = 0.07 \text{ IR for the 1-ft pipeline}$$

(continued)

Now, add the inactivation ratios for summer and winter.

$$\text{Summer} = 3.2 + 0.04 + 0.03 = 3.27, \text{ round to } \mathbf{3.3\ IR}$$

This meets the CTs required.

$$\text{Winter} = 4.72 + 0.09 + 0.07 = 4.48\ \text{IR, round to } \mathbf{4.9\ IR}$$

This meets the CTs required.

116. Calculate the CT for a direct filtration plant with the following characteristics to determine if it meets the Surface Water Treatment Rule (SWTR):

Clear well capacity = 3.0 mil gal

pH = 7.6

Flow = 8.5 mgd

Temperature = 11°C

18-in. pipeline is 785 ft to first customer.

Chlorine residual at outlet of clear well is 1.2 mg/L.

Chlorine residual at first customer is 0.6 mg/L and pH = 7.5.

Clear well has poor baffling and tracer studies indicate T_{10} = 40 min.

First, the SWTR requires a 3 log removal of *Giardia* cysts. From Table 4-6, a plant with direct filtration has 2 log credits. Thus, we need at least 1.0 log removal to be in compliance.

A tracer study was done to determine the detention time and it has a value of 30 min. Also, because a tracer study has been done, the conversion factor can be disregarded. (***Note***: Poor baffling of the clear well does not matter in this problem.)

Then, calculate the CT:

$$CT_{calc} = (\text{Conversion factor})(\text{Detention time, in min})(Cl_2 \text{ residual})$$

$$CT_{calc} = (40 \text{ min})(1.2 \text{ mg/L}) = 48\ CT_{calc}$$

From the disk provided, find the CT_{req}. Find the log inactivation table of 1.0 with a pH of 7.6. Go to 11°C and the chlorine residual of 1.2 mg/L. In this case it is 44.7.

$$\textbf{Inactivation ratio (IR)} = \mathbf{CT_{calc}/\ CT_{req}}$$

$$= 48/44.7 = 1.07 \text{ IR for the clear well}$$

This indicates that the plant is in compliance.

We need to determine the CT for the pipeline.

First, find the diameter in feet for the 18-in. pipeline.

$$\text{Diameter} = \frac{(18 \text{ in.})(1 \text{ ft})}{12 \text{ in.}} = 1.5 \text{ ft}$$

$$\textbf{Volume} = \textbf{(0.785)(Diameter)}^2\textbf{(Length)}$$

Next, find the radius.

$$\text{Volume} = (0.785)(1.5 \text{ ft})(1.5 \text{ ft})(785 \text{ ft}) = 1{,}386.5 \text{ ft}^3$$

$$\text{Number of gallons} = (1{,}386.5 \text{ ft}^3)(7.48 \text{ gal/ft}^3) = 10{,}371 \text{ gal}$$

$$\text{Detention time} = \text{Volume/Flow} = \frac{10{,}371 \text{ gal } (1{,}440 \text{ min/d})}{8{,}500{,}000 \text{ gpd}} = 1.76 \text{ min}$$

$$CT_{calc} = (1.0 \text{ conversion factor})(1.76 \text{ min})(0.6 \text{ mg/L}) = 1.06\ CT_{calc}$$

From the disk provided, find the CT_{req}. Temperature is 11°C and chlorine residual is 0.6 mg/L. In this case, it is 40.2.

$$IR = CT_{calc}/\ CT_{req} = 1.06/40.2 = 0.026 \text{ IR for the pipeline}$$

Now, add the two inactivation ratios (clear well and pipeline).

1.07 + 0.026 = 1.096 IR, round to **1.1 IR**

This plant meets the required CTs for removal of *Giardia* cysts.

117. Calculate the CT for a slow sand filtration plant with the following characteristics to determine if it meets the Surface Water Treatment Rule (SWTR).

Clear well capacity = 4 mil gal with poor baffling

pH = 8.0

Flow = 1.1 mgd

Temperature = 6°C

24-in. pipeline is 362 ft to first customer.

Chlorine residual at outlet of clear well is 1.0 mg/L.

Chlorine residual at first customer is 0.8 mg/L.

Clear well has no baffling and tracer studies indicate T_{10} = 58 min.

(continued)

First, the SWTR requires a 3 log removal of *Giardia* cysts. From Table 4-6, a plant with slow sand filtration has 2 log credits. Thus, we need at least 1.0 log removal to be in compliance.

Calculate the CT.

$$CT_{calc} = \text{(Conversion factor)(Detention time, in min)}(Cl_2 \text{ residual})$$

The tracer study gives the detention time, and it has a value of 58 min (eliminates conversion factor).

$$CT_{calc} = (58 \text{ min})(1.0 \text{ mg/L}) = 58\ CT_{calc}$$

On the disk included in this book, find the CT_{req}. Find the log inactivation table of 1.0 with a pH of 8.0. Now, go to 6°C and the chlorine residual of 1.0 mg/L. In this case, it is 68.4.

$$\textbf{Inactivation ratio (IR)} = \mathbf{CT_{calc}/CT_{req}}$$

$$= 58/68.4 = 0.85 \text{ IR for the clear well}$$

Now, determine the CT for the pipeline.

First, find the volume of the pipeline.

The equation is:

$$\textbf{Volume} = \mathbf{(0.785)(Diameter)^2(Length)}$$

$$\text{Volume} = (0.785)(2.0 \text{ ft})(2.0 \text{ ft})(362 \text{ ft}) = 1{,}137 \text{ ft}^3$$

Then, find the number of gallons.

$$\text{Number of gallons} = (1{,}137 \text{ ft}^3)(7.48 \text{ gal/ft}^3) = 8{,}505 \text{ gal}$$

$$\text{Detention time} = \text{Volume/Flow} = (8{,}505 \text{ gal})(1{,}440 \text{ min/d})/1{,}100{,}000 \text{ gpd} = 11.1 \text{ min}$$

$$CT_{calc} = (1.0)(11.1 \text{ min})(0.8 \text{ mg/L}) = 8.88\ CT_{calc}$$

From the disk provided, find the CT_{req}. Find the log inactivation table of 1.0 with a pH of 8.0. Go to 6°C and the chlorine residual of 0.8 mg/L. In this case it is 66.6.

$$\text{IR} = CT_{calc}/\ CT_{req} = 8.88/66.6 = 0.13 \text{ IR for the pipeline}$$

Now, add the two inactivation ratios (clear well and pipeline).

$$0.85 + 0.13 = \mathbf{0.98\ IR}$$

This plant does not meet the 1.0 CT criteria required for the removal of *Giardia* cysts.

BLENDING CALCULATIONS

The problems in this section use parts per million. Remember that parts per million is the same as milligrams per liter.

118. Determine the nitrate blending for two wells given the following data. Find the maximum flow that will give a nitrate level of less than 35 ppm. Give result to the nearest gallon.

Well 1		Well 2	
Flow, gpm	**Nitrate, ppm**	**Flow, gpm**	**Nitrate, ppm**
280	82	75	11
250	80	70	10
225	81	60	9
200	84	65	9
175	78	80	11
150	79	60	8

The equation is:

$$\mathbf{Q_1C_1 + Q_2C_2 = Q_3C_3}$$

where: Q_1 = Flow from well 1 (the unknown)
C_1 = Highest concentration from well 1
Q_2 = Lowest flow from well 2
C_2 = Highest nitrate value from well 2
$Q_3 = Q_1 + Q_2$
C_3 = Highest nitrate level permissible (35 ppm)

Solve for Q_1: Do multiplication first.

$$Q_1\text{ (84 ppm)} + \text{(60 gpm)(11 ppm)} = (Q_1 + \text{60 gpm})\text{(35 ppm)}$$

Subtract Q_1(35 ppm) from each side of the equation.

$$\begin{array}{lcl} Q_1\text{ (84 ppm)} + \text{(660 gpm ppm)} & = & Q_1\text{ (35 ppm)} + \text{2,100 gpm ppm} \\ -Q_1\text{ (35 ppm)} & & -Q_1\text{ (35 ppm)} + \text{2,100 gpm} \\ \hline Q_1\text{ (49 ppm)} + \text{(660 gpm ppm)} & = & \text{2,100 ppm} \end{array}$$

Subtract 660 gpm ppm from each side of the equation.

$$Q_1\text{(49 ppm)} = \text{1,440 gpm ppm}$$

$$Q_1 = \textbf{29 gpm maximum flow from well 1}$$

119. Determine the nitrate blending for two wells given the following data. Find the maximum flow that will give a nitrate level of less than 35 ppm. Give result to the nearest gallon.

Well 3A		Well 4B	
Flow, gpm	**Nitrate, ppm**	**Flow, gpm**	**Nitrate, ppm**
450	109	520	21
430	102	500	20
465	118	525	21
440	106	545	19
425	103	530	21
445	101	510	18

The equation is:

$$Q_1C_1 + Q_2C_2 = Q_3C_3$$

where: Q_1 = Flow from well 3A (the unknown)
C_1 = Highest concentration from well 3A
Q_2 = Lowest flow from well 4B
C_2 = Highest nitrate value from well 4B
$Q_3 = Q_1 + Q_2$
C_3 = Highest nitrate level permissible (35 ppm)

Solve for Q_1: Do multiplication first.

$$Q_1\ (118\text{ ppm}) + (500\text{ gpm})(21\text{ ppm}) = (Q_1 + 500\text{ gpm})(35\text{ ppm})$$

Subtract Q_1(35 ppm) from each side of the equation.

$$\begin{array}{lcr} Q_1\ (118\text{ ppm}) + (10{,}500\text{ gpm ppm}) & = & Q_1\ (35\text{ ppm}) + 17{,}500\text{ gpm ppm} \\ -Q_1\ (\ 35\text{ ppm}) & & -Q_1\ (35\text{ ppm}) \\ \hline Q_1\ (\ 83\text{ ppm}) + (10{,}500\text{ gpm ppm}) & = & 17{,}500\text{ gpm ppm} \end{array}$$

Subtract 10,500 gpm ppm from each side of the equation.

$$Q_1(83\text{ ppm}) = 7{,}000\text{ gpm ppm}$$

Q_1 = **84 gpm maximum flow from well 3A**

120. Determine the iron blending for three wells given the following data. Find the maximum flow that will give an iron level of less than 0.03 ppm. Give the result to the nearest gallon.

Well D-3		**Well D-5**		**Well C-1**	
Flow	**Fe**	**Flow**	**Fe**	**Flow**	**Fe**
360	**0.015**	**120**	**0.061**	**210**	**0.107**
325	**0.013**	**150**	**0.065**	**225**	**0.105**
380	**0.015**	**130**	**0.062**	**230**	**0.106**
375	**0.014**	**115**	**0.061**	**240**	**0.104**
350	**0.016**	**100**	**0.065**	**250**	**0.103**
340	**0.012**	**140**	**0.064**	**260**	**0.101**

The equation is:

$$\mathbf{Q_1C_1 + Q_2C_2 + Q_3C_3 = Q_4C_4}$$

where: Q_1 = Flow from well C-1 (the unknown)
C_1 = Highest Fe concentration from well C-1
Q_2 = Lowest flow from well D-3
C_2 = Highest Fe concentration from well D-3
Q_3 = Lowest flow from well D-5
C_3 = Highest Fe concentration from well D-5
$Q_4 = Q_1 + Q_2 + Q_3$
C_4 = Highest Fe level desired (0.03 ppm)

$$Q_1\text{ (0.107 ppm)} + \text{(325 gpm)(0.016 ppm)} + \text{(100 gpm)(0.065 ppm)} = (Q_1 + \text{325 gpm} + \text{100 gpm)(0.03 ppm)}$$

First, multiply products.

$$Q_1\text{ (0.107 ppm)} + \text{(5.2 gpm ppm)} + \text{(6.5 gpm ppm)} = 0.03Q_1 + \text{9.75 gpm ppm} + \text{3 gpm ppm}$$

Add like quantities and then subtract Q_1(0.03 ppm) from each side of the equation:

$$\begin{aligned} Q_1\text{ (0.107 ppm)} + \text{11.7 gpm ppm} &= 0.03Q_1 + \text{12.75 gpm ppm} \\ Q_1\text{ (0.107 ppm)} + \text{(11.7 gpm ppm)} &= Q_1\text{(0.03 ppm)} + \text{12.75 gpm ppm} \\ Q_1\text{ (0.03 ppm)} & \quad -Q_1\text{(0.03 ppm)} \\ \hline Q_1\text{ (0.077 ppm)} + \text{(11.7 gpm ppm)} &= \text{12.75 gpm ppm} \end{aligned}$$

Subtract 11.7 gpm ppm from each side of the equation

$$Q_1\text{(0.077 ppm)} = \text{1.05 gpm ppm}$$

$$Q_1 = \frac{\text{1.05 gpm ppm}}{\text{0.077 ppm}}$$

Q_1 = 13.64 gpm, round to **14 gpm maximum flow from well C-1**

121. In the previous problem, what would be the maximum flow in well D-5 to give an iron level of less than 0.03 ppm if well C-1 were not used?

Well D-3		Well D-5	
Flow, gpm	**Fe, ppm**	**Flow, gpm**	**Fe, ppm**
360	**0.015**	**120**	**0.061**
325	**0.013**	**150**	**0.066**
380	**0.015**	**130**	**0.062**
375	**0.014**	**115**	**0.061**
350	**0.016**	**100**	**0.065**
340	**0.012**	**140**	**0.064**

The equation is:

$$\mathbf{Q_1C_1 + Q_2C_2 = Q_3C_3}$$

where: Q_1 = Flow from well D-5 (the unknown)
C_1 = Highest concentration from well D-5
Q_2 = Lowest flow from well D-3
C_2 = Highest iron value from well D-3
$Q_3 = Q_1 + Q_2$
C_3 = Highest iron level desired (0.03 ppm)

Solve for Q_1: Do multiplication first.

$$Q_1\ (0.066\text{ ppm}) + (325\text{ gpm})(0.016\text{ ppm}) = (Q_1 + 325\text{ gpm})(0.03\text{ ppm})$$

Subtract Q_1(0.03 ppm) from each side of the equation.

$$\begin{array}{ll} Q_1\ (0.066\text{ ppm}) + (5.2\text{ gpm ppm}) & = Q_1(0.03\text{ ppm}) + 9.75\text{ gpm ppm} \\ Q_1\ (0.03\text{ ppm}) & \quad - Q_1(0.03\text{ ppm}) \\ \hline Q_1(0.036\text{ ppm}) + (5.2\text{ gpm ppm}) & = 9.75\text{ gpm ppm} \end{array}$$

Subtract 5.2 gpm ppm from each side of the equation.

$$Q_1(0.036\text{ ppm}) = 4.55\text{ gpm ppm}$$

$$Q_1 = \mathbf{126\text{ gpm}}$$

There are only two significant figures for this problem. It is best to round down to make sure iron does not exceed 0.3 mg/L. Thus, 120 gpm maximum flow from well D-5 is the best answer.

122. Determine the nitrate blending for two wells given the following data. Find the maximum flow that will give a nitrate level of less than 35 mg/L. Give the result to the nearest gallon.

Well CP3P		Well CP3Q	
Flow, gpm	**Nitrate, ppm**	**Flow, gpm**	**Nitrate, ppm**
280	85	320	22
270	85	300	19
290	83	310	20
295	81	295	19
285	83	290	18
260	88	280	19

The equation is:

$$Q_1C_1 + Q_2C_2 = Q_3C_3$$

where: Q_1 = Flow from well CP3P (the unknown)
C_1 = Highest concentration from well CP3P
Q_2 = Lowest flow from well CP3Q
C_2 = Highest nitrate value from well CP3Q
$Q_3 = Q_1 + Q_2$
C_3 = Highest nitrate level permissible (35 mg/L)

Solve for Q_1: Do multiplication first.

$$Q_1 \text{ (88 ppm)} + \text{(280 gpm)(22 ppm)} = (Q_1 + \text{280 gpm)(35 ppm)}$$

Subtract Q_1(35 ppm) from each side of the equation.

$$\begin{array}{lcr} Q_1 \text{ (88 ppm)} + \text{(6,160 gpm ppm)} & = & Q_1\text{(35 ppm)} + \text{9,800 gpm ppm} \\ Q_1 \text{ (35 ppm)} & & -\,Q_1\text{(35 ppm)} \\ \hline Q_1 \text{ (53 ppm)} + \text{(6,160 gpm ppm)} & = & \text{9,800 gpm ppm} \end{array}$$

Subtract 6,160 gpm ppm from each side of the equation.

$$Q_1\text{(53 ppm)} = \text{3,640 gpm ppm}$$

Q_1 = 68.7, round down to be safe to **68 gpm maximum flow from well CP3P**

123. Determine the nitrate blending for two wells given the following data. Find the maximum flow that will give a nitrate level of less than 35 ppm.

Well 1		Well 2	
Flow, gpm	**Nitrate, ppm**	**Flow, gpm**	**Nitrate, ppm**
120	67	90	19
140	70	95	22
145	74	85	21
135	68	80	19
130	62	87	20
115	65	92	22

The equation is:

$$Q_1C_1 + Q_2C_2 = Q_3C_3$$

where: Q_1 = Flow from well 1 (the unknown)
C_1 = Highest concentration from well 1
Q_2 = Lowest flow from well 2
C_2 = Highest nitrate value from well 2
$Q_3 = Q_1 + Q_2$
C_3 = Highest nitrate level permissible (35 ppm)

Solve for Q_1: Do multiplication first.

$$Q_1 \text{ (74 ppm)} + \text{(80 gpm)(22 ppm)} = (Q_1 + \text{80 gpm})\text{(35 ppm)}$$

Subtract Q_1(35 ppm) from each side of the equation.

$$\begin{array}{ll} Q_1 \text{ (74 ppm)} + \text{(1,760 gpm ppm)} = & Q_1\text{(35 ppm)} + \text{2,800 gpm ppm} \\ Q_1 \text{ (35 ppm)} & -\,Q_1\text{(35 ppm)} \\ \hline Q_1\text{(39 ppm)} + \text{(1,760 gpm ppm)} = & \text{2,800 gpm ppm} \end{array}$$

Subtract 1,760 gpm ppm from each side of the equation.

$$Q_1\text{(39 ppm)} = \text{1,040 gpm ppm}$$

Divide each side by 39 ppm.

Q_1 = **27 gpm maximum flow from well 1**

PREPARATION OF STANDARD SOLUTIONS

124. A 0.2 normal solution of NaOH is to be prepared. If 1 L of solution is desired, how many grams of NaOH are required? The gram formula for NaOH is 40.00.

$$\textbf{Normality (N)} = \frac{\textbf{Number of gram-equivalents of solute}}{\textbf{Number of liters of solution}}$$

Substituting:

$$0.2\text{ N} = \frac{\text{Number of gram-equivalents of solute}}{1\text{ L of solution}} = 0.2\text{ gram-equivalents are required}$$

The formula weight for NaOH = gram equivalent weight in this case.

Number of grams = (Number of gram-equivalents)(Number grams/gram-equivalents)

Substituting:

(0.2 gram-equivalents)(40.00 grams/gram-equivalents)

= **8 g dissolved in 1 L of deionized water**

125. To prepare a 0.1 normal solution of NaCl (sodium chloride or table salt), 5.844 g are required. If 5.872 g are weighed out, how many milliliters of deionized water should be used?

The equation is:

$$\frac{\textbf{Desired weight, grams}}{\textbf{Desired solution volume, mL}} = \frac{\textbf{Actual weight, grams}}{\textbf{Actual solution volume, mL}}$$

Substituting:

$$\frac{5.844\text{ g}}{1{,}000\text{ mL}} = \frac{5.872\text{ g}}{x\text{, mL}}$$

Solve for x.

$$x\text{, mL} = \frac{(1{,}000\text{ mL})(5.872\text{ g})}{5.844\text{ g}}$$

$$x\text{, mL} = \mathbf{1{,}005\text{ mL}}$$

Note: It is much easier to weigh out a little less than what is desired so that you do not need more than 1,000 mL to make up the solution. Why? Precision glassware, such as a volumetric flask, comes in 1,000 mL, not 1,005 mL.

126. A 0.25 normal copper sulfate ($CuSO_4$) solution is to be made using 39.9 g. If 39.67 g were weighed out, how many milliliters of deionized water should be used?

The equation is:

$$\frac{\textbf{Desired weight, grams}}{\textbf{Desired solution volume, mL}} = \frac{\textbf{Actual weight, grams}}{\textbf{Actual solution volume, mL}}$$

$$\frac{39.9\text{ g}}{1{,}000\text{ mL}} = \frac{39.67\text{ g}}{x\text{, mL}}$$

$$x\text{, mL} = \frac{(1{,}000\text{ mL})(39.67\text{ g})}{39.9\text{ g}}$$

$$x\text{, mL} = \textbf{994 mL of deionized water}$$

To make the 0.25*N* $CuSO_4$ solution, fill a 1,000-mL volumetric flask to the 1,000-mL mark with deionized water. Remove 6 mL using a pipette. Pour this deionized water into a mixing container. Add the 39.67 g of $CuSO_4$ and mix. ***Note***: In this case it would be much easier to measure out exactly 39.9 g of $CuSO_4$.

127. How many milliliters of a 1.0 standard solution of hydrochloric acid (HCl) will neutralize 100 mL of a 2.0 normal sodium hydroxide (NaOH) solution?

Because the chemicals react in a 1:1 ratio, the equation is:

$$\textbf{(Concentration}_1\textbf{)(Volume}_1\textbf{)} = \textbf{(Concentration}_2\textbf{)(Volume}_2\textbf{)}$$

$$(\text{mL HCl})(\text{N HCl}) = (\text{mL NaOH})(\text{N NaOH})$$

$$(\text{mL HCl})(1.0\text{ N HCl}) = (100\text{ mL NaOH})(2.0\text{ N NaOH})$$

Solve for mL HCl.

$$\text{mL HCl} = \frac{(100\text{ mL NaOH})(2.0\text{ N NaOH})}{1.0\text{ N HCl}} = \textbf{200 mL}$$

128. A 0.50 normal solution of calcium carbonate ($CaCO_3$) is to be prepared. If 1 L of solution is desired, how many grams of $CaCO_3$ are required? The gram formula for $CaCO_3$ is 100.09.

$$\textbf{Normality (N)} = \frac{\textbf{Number of gram-equivalents of solute}}{\textbf{Number of liters of solution}}$$

$$0.50\text{ N} = \frac{\text{Number of gram-equivalents of solute}}{\text{1 L of solution}} = 0.50\text{ gram-equivalents are required}$$

The formula weight for $CaCO_3$ = gram equivalent weight.

Number of grams = (Number of gram-equivalents)(Number grams/gram-equivalents)

(0.50 gram-equivalents)(100.09 grams/gram-equivalents)

= **50 g dissolved in 1 L of deionized water**

129. A 0.20 normal solution of phosphoric acid (H_3PO_4) is to be prepared. If 0.5 L of solution is desired, how many grams of H_3PO_4 are required? The gram formula for H_3PO_4 is 98.00.

$$\textbf{Normality (N)} = \frac{\textbf{Number of gram-equivalents of solute}}{\textbf{Number of liters of solution}}$$

$$0.20\text{ N} = \frac{\text{Number of gram-equivalents of solute}}{\text{1 L of solution}} = 0.20\text{ gram-equivalents are required}$$

The formula weight for H_3PO_4 = ⅓ the gram equivalent weight because 3 hydrogen atoms (H^+) combine with 1 PO_4^{-3}. Therefore, divide the gram formula weight by 3 (98/3 = 32.67), or 32.67 grams per equivalent weight.

Number of grams = (Number of gram-equivalents)(Number grams/gram-equivalents)

(0.20 gram-equivalents)(32.67 grams/gram-equivalents)

= **6.5 g dissolved in 1 L of deionized water**

130. What is the normality of a HCl solution that has 1.5 equivalents in 835 mL?

First, convert the number of milliliters to liters.

$$835 \text{ mL}/1{,}000 \text{ mL per L} = 0.835 \text{ L}$$

Then, find the normality of the solution.

$$\text{Normality (N)} = \frac{\text{Number of gram-equivalents of solute}}{\text{Number of liters of solution}} = \frac{1.5 \text{ gram-equivalents}}{0.835 \text{ L}}$$

$$= \mathbf{1.8\ gram\text{-}equivalents/L,\ or\ 1.8\ N}$$

131. What is the concentration of alum in milligrams per liter if 26 mL of a 0.45-g/L alum solution is added to 1,000 mL of deionized water?

The equation is:

$$\textbf{Alum dosage, mg/L} = \frac{\textbf{(Stock, mL)(1,000 mg/g)(Concentration in grams/liter)}}{\textbf{Sample size, mL}}$$

$$\text{Alum dosage, mg/L} = \frac{(26 \text{ mL})(1{,}000 \text{ mg/g})(0.45 \text{ g/L})}{1{,}000 \text{ mL}} = \mathbf{12\ mg/L}$$

132. What is the normality (N) of a NaOH solution that has 1.35 equivalents in 450 mL?

First, convert the number of milliliters to liters.

$$450 \text{ mL}/1{,}000 \text{ mL per L} = 0.45 \text{ L}$$

Then, find the normality of the solution.

$$\text{N} = \frac{\text{Number of gram-equivalents of solute}}{\text{Number of liters of solution}} = \frac{1.35 \text{ equivalents}}{0.45 \text{ L}} = \mathbf{3.0\ N}$$

JAR TEST STANDARDS

133. A 0.1% standard solution (1,000 ppm or 1,000 mg/L) is required for doing jar tests. If the alum has a specific gravity of 1.25 and is 48.0% alum, how many milliliters of alum are required to make exactly 1,000 mL of stock solution?

First, find the number of pounds per gallon of alum.

$$\text{Pounds per gallon, alum} = (\text{Specific gravity})(8.34 \text{ lb/gal}) = (1.25)(8.34 \text{ lb/gal}) = 10.425 \text{ lb/gal}$$

Next, determine the number of grams per milliliter.

$$\text{Number of g/mL, alum} = \frac{(10.425 \text{ lb/gal})(48.0\% \ Al_2SO_4 \text{ purity})(454 \text{ g/lb})}{(3{,}785 \text{ mL/gal})(100\%)}$$

$$\text{Number of g/mL, alum} = 0.600 \text{ g/mL}$$

Convert grams per milliliter to milligrams per milliliter.

$$\text{Number of mg/mL} = (0.600 \text{ g/mL})(1{,}000 \text{ mg/g}) = 600 \text{ mg/mL}$$

Next, convert milliliters to liters by multiplying by 1,000.

$$\text{Number of mg/L} = (600 \text{ mg/mL})(1{,}000 \text{ mL/L}) = 600{,}000 \text{ mg/L}$$

Next, determine the number of milliliters required.

The equation: $C_1V_1 = C_2V_2$

$$(600{,}000 \text{ mg/mL})(x, \text{ mL}) = (1{,}000 \text{ mg/L})(1{,}000 \text{ mL})$$

$$x, \text{ mL} = \frac{(1{,}000 \text{ mg/L})(1{,}000 \text{ mL})}{600{,}000 \text{ mg/L}} = 1.67 \text{ mL, alum}$$

Now, using a micropipette, add 1.67 mL of the 48.0 percent alum solution to a clean, dry 1,000-mL flask. Dilute the alum to the 1,000-mL mark with deionized water. Add a magnetic stir bar and place the flask on a magnetic stirrer. Turn the magnetic stirrer on and mix this solution with the bar as vigorously as possible for at least 10 minutes. Thus, every 1 mL of alum solution that is added to a 1,000-mL raw water sample will add a 1-mg/L dose. If 10 mL of this stock solution were added, it would be a dose of 10 mg/L. If you are using the 2-L square jars, simply double the milliliters added for each mg/L dosage desired. Another way is to feed the alum neat by using a micropipette; pipette 0.00167 mL of alum into a 1,000-mL raw water sample or, if you are using 2-L square beakers, simply double this amount.

134. A 1,000 mg/L polymer solution (1,000 ppm or 1,000 mg/L) is desired for performing a jar test. If the polymer has a specific gravity of 1.33 and is 100% polymer, how many milliliters of polymer are required to make exactly 1,000 mL of stock solution?

First, find the number of pounds per gallon of polymer.

$$\text{Pounds per gallon, polymer} = (\text{Specific gravity})(8.34 \text{ lb/gal}) = (1.33)(8.34 \text{ lb/gal}) = 11.0922 \text{ lb/gal}$$

Next, determine the number of grams per milliliter.

(continued)

$$\text{Number of g/mL, polymer} = \frac{(11.0922\text{ lb/gal})(100\%\text{ polymer})(454\text{ g/lb})}{(3{,}785\text{ mL/gal})(100\%)}$$

$$\text{Number of g/mL, polymer} = 1.33\text{ g/mL}$$

Convert grams per milliliter to milligrams per milliliter.

$$\text{Number of mg/mL} = (1.33\text{ g/mL})(1{,}000\text{ mg/g}) = 1{,}330\text{ mg/mL}$$

Next, convert milliliters to liters by multiplying by 1,000.

$$\text{Number of mg/L} = (1{,}330\text{ mg/mL})(1{,}000\text{ mL/L}) = 1{,}330{,}000\text{ mg/L}$$

Next, determine the number of milliliters required.

The equation: $\mathbf{C_1V_1 = C_2V_2}$

$$(1{,}330{,}000\text{ mg/L})(x,\text{ mL}) = (1{,}000\text{ mg/L})(1{,}000\text{ mL})$$

$$x,\text{ mL} = \frac{(1{,}000\text{ mg/L})(1{,}000\text{ mL})}{1{,}330{,}000\text{ mg/L}} = \mathbf{0.75\text{ mL, polymer}}$$

Now, using a micropipette, add 0.75 mL of the 100 percent polymer solution to a clean, dry 1,000-mL flask. Dilute the polymer to the 1,000-mL mark with deionized water. Add a magnetic stir bar and place the flask on a magnetic stirrer. Turn the magnetic stirrer on and mix this solution with the bar as vigorously as possible for at least 10 minutes. Thus, every 1 mL of polymer solution that is added to a 1,000-mL raw water sample will add a 1-mg/L dose. If 10 mL of this stock solution were added, it would be a dose of 10 mg/L. If you are using the 2-L square jars, simply double the milliliters added for each mg/L dosage desired. Another way is to feed the polymer neat by using a micropipette; pipette 0.00075 mL of polymer into a 1,000-mL raw water sample or, if you are using 2-L square beakers, simply double this amount.

135. A solution of lime needs to be prepared for a jar test. How many grams of quicklime, CaO, would you mix with 1 L of water to make a 0.25% (weight-volume) solution?

The equation for weight-volume percent is:

$$\textbf{Percent of solution} = \frac{\textbf{Weight of solute, g}}{\textbf{Volume of solution, mL}} \times \mathbf{100\%}$$

First, determine the volume in milliliters. Note that 1 L = 1,000 mL.

$$0.25\% = \frac{\text{Weight of solute, g}}{1{,}000\text{ mL}} \times 100\%$$

Solve for weight of solute, grams, by multiplying both sides of the equation by 1,000 mL and dividing both sides by 100%.

Thus,

$$\text{Weight of solute, g} = \frac{(0.25\%)(1{,}000\text{ mL})}{100\%} = 2.5\text{ g of CaO dissolved in 1 L of water}$$

136. A solution of lime needs to be prepared for a jar test. How many grams of quicklime, CaO, would you mix with 1 L of water to make a 0.05% (weight-volume) solution?

The equation is:

$$\textbf{Percent of solution} = \frac{\textbf{Weight of solute, g}}{\textbf{Volume of solution, mL}} \times \textbf{100\%}$$

First, determine the volume in milliliters. Note that 1 L = 1,000 mL.

$$0.05\% = \frac{\text{Weight of solute, g}}{1{,}000 \text{ mL}} \times 100\%$$

Solve for weight of solute, grams, by multiplying both sides of the equation by 1,000 mL and dividing both sides by 100%.

Thus,

$$\text{Weight of solute, g} = \frac{(0.05\%)(1{,}000 \text{ mL})}{100\%} = \textbf{0.5 g of CaO dissolved in 1 L of water}$$

DILUTION PROBLEMS

137. What is the concentration of alum, in milligrams per liter, if 15 mL of a 0.20-g/L alum solution is added to 1,000 mL of deionized water?

The equation is:

$$\textbf{Alum dosage, mg/L} = \frac{\textbf{(Stock, mL)(1,000 mg/g)(Concentration, g/L)}}{\textbf{Sample size, mL}}$$

$$\text{Alum dosage, mg/L} = \frac{(15 \text{ mL})(1{,}000 \text{ mg/g})(0.20 \text{ g/L})}{1{,}000 \text{ mL}} = \textbf{3.0 mg/L}$$

138. A procedure calls for diluting a 70% potassium iodide (KI) solution to a 20% solution. If exactly 1,000 mL of the 20% solution is desired, how much of the 70% solution is required. The atomic weight of K = 39.102 g/mol and for I = 129.9044 g/mol, or a gram formula weight of KI (rounded to the nearest hundredth of a gram) is 166.01 g.

Solve using the dilution triangle.

70%		20	Use 20 parts of the 70% solution for every 70 parts
	20%		
0%		50	Use 50 parts of the dilution water (deionized water) for every 70 parts
		70 parts	

$$\frac{(20 \text{ parts})(1{,}000 \text{ mL})}{70 \text{ part}} = \textbf{286 mL of the 70\% KI solution}$$

$$\frac{(50 \text{ parts})(1{,}000 \text{ mL})}{70 \text{ parts}} = \frac{\textbf{714 mL of the deionized (DI) water}}{1{,}000 \text{ mL}}$$

Mix 286 mL of the 70% KI solution in 714 mL of DI water.

Another way to solve this problem is as follows.

The equation is:

$$C_1V_1 + C_2V_2 = C_3V_3$$

where: C_1 = the concentration and V_1 is the volume of the 70% KI solution

C_2 = the concentration and V_2 is the volume of the deionized water used to dilute the KI solution

C_3 = the final concentration of the mixture and V_3 is the final volume of the mixture

Because V_1 and V_2 are unknown, set $V_1 = V_3 - V_2$, since volume 1 + volume 2 has to equal volume 3.

That is, $V_1 + V_2 = V_3$. Subtracting V_2 from both sides of the equation gives $V_1 = V_3 - V_2$.

Substituting:

$$(70\%/100\%)(1{,}000 \text{ mL} - V_2) + (0\%)(V_2) = (20\%/100\%)(1{,}000 \text{ mL})$$

$$700 \text{ mL} - 0.7V_2 + 0 = 200 \text{ mL}$$

Subtract 200 mL from both sides of the equation.

$$500 \text{ mL} - 0.7V_2 = 0$$

Subtract 500 mL from both sides of the equation.

$$-0.7V_2 = -500 \text{ mL}$$

Multiply both sides of the equation by (–1)

$$0.7V_2 = 500 \text{ mL}$$

Divide both sides of the equation by 0.7.

V_2 = **714 mL of deionized water needed to dilute the KI solution of 70% to 20%**

$$V_1 = V_3 - V_2$$

V_1 = 1,000 mL – 714 mL = **286 mL of the 70% KI solution**

This solution is more complicated, but if you prefer, use. it.

MILLIAMP SIGNAL CALCULATIONS

139. A storage tank has a capacity of 30 ft. Currently there are 18.8 ft of water in the tank. What would the SCADA reading be on the board in milliamps (mA) for a 4-mA to 20-mA signal?

Equation:

$$\textbf{Current process reading} = \frac{\textbf{(Live signal, mA – 4 mA offset)(Maximum capacity)}}{\textbf{16 milliamp span}}$$

Substitute known values and solve:

$$18.8 \text{ ft (Storage tank level)} = \frac{\text{(Live signal mA – 4 mA offset)(30 ft maximum level)}}{16 \text{ mA}}$$

Rearrange the equation to solve for live signal in mA.

$$\text{Live signal mA} - 4 \text{ mA offset} = \frac{(18.8 \text{ ft})(16 \text{ mA})}{30 \text{ ft}}$$

$$\text{Live signal mA} = \frac{(18.8 \text{ ft})(16 \text{ mA})}{30 \text{ ft}} + 4 \text{ mA offset}$$

$$\text{Live signal mA} = 14.03 \text{ mA, round to 14 mA}$$

140. The SCADA system at a water plant uses a 4 mA to 20 mA signal to monitor tank levels. If the readout on the SCADA board reads 15 mA, what is the height of the water in a tank with a capacity of 28 ft?

Note that 4 mA = 0 ft in the tank and that 20 mA = 28 ft in the tank.

Based on this, first subtract 4 mA from the board reading and from the maximum amount as follows:

$$15 \text{ mA} - 4 \text{ mA} = 11 \text{ mA}$$

$$20 \text{ mA} - 4 \text{ mA} = 16 \text{ mA}$$

Next, divide the current board reading by the maximum reading.

$$\frac{11 \text{ mA}}{16 \text{ mA}} = 0.6875$$

Each foot in the tank will cause a milliamp reading of 0.6875.

Height of water in the tank is equal to this ratio multiplied by the maximum height of water in feet.

$$\text{Height, ft} = (0.6875)(28 \text{ ft}) = \mathbf{19.25 \text{ ft}}$$

Water Treatment Grade 4 Test

(answers on p. 326)

1. Exactly 60 gal of a 4.5% hypochlorite solution is needed. How many gallons of a 9.2% solution must be mixed with a 1.1% solution to make the required solution?

2. What is the percent potassium (K) in potassium permanganate ($KMnO_4$)?

3. Water is flowing at a velocity of 1.35 ft/s in a 10.0-in. diameter pipe. If the pipe changes from 10.0 in. to 6.0 in., what will the velocity be in the 6.0-in. pipe?

4. A storage tank contains 235 gal of a 14.6% zinc orthophosphate (ZOP) solution that weighs 10.21 lb/gal. Note that 1,500 gal of a 11.9% ZOP solution is added to the tank. Paperwork on the new ZOP indicates that the specific gravity is 1.17 lb/gal. What is the percent strength of the resulting solution?

5. Find the amount of iron (Fe) and manganese (Mn) removed per year from a plant that treats an average of 38 mgd if the average iron concentration is 1.58 ppm and the average manganese concentration is 0.35 ppm. The removal efficiency is 89.1% for the iron and 75.2% for the manganese.

6. A 4-min drawdown test result shows that 171 mL of a cationic polymer is being used to treat the raw water. The specific gravity of the polymer is 1.37. If the plant is treating 7,250 gpm, what is the polymer dosage in milligrams per liter?

7. The level in a storage tank rises 4.27 ft in 5.5 hr. If the tank has a diameter of 225 ft and the plant is producing 23.9 mgd, what is the average discharge rate of the treated water discharge pumps in gallons per minute?

8. A softener unit has 575 ft^3 of resin with a capacity of 28.1 kilograins/ft^3. How many gallons of water will the unit treat if the water contains 19.6 gpg?

9. Determine the feed rate for alum in milliliters per minute under the following conditions:

Plant flow	17.5 mgd
Alum dosage rate	15.6 mg/L
Alum percentage	48.0%
Alum specific gravity	1.32

10. A water treatment plant has a flow of 46 mgd and is being treated with 1,985 gpd of a hypochlorite solution. If the desired dose is 3.0 mg/L, determine the concentration of the hypochlorite solution, in percent. The hypochlorite solution weighs 9.25 lb/gal. *Note*: For simplicity, assume no chlorine demand.

Solutions to Water Treatment Grade 4 Test

1. Solve the problem using the dilution triangle.

9.2%	3.4	3.4 parts of the 9.2% solution are required for every 8.1 parts
4.5%		
1.1%	4.7	4.7 parts of the 1.1% solution are required for every 8.1 parts
	8.1 total parts	

$$\frac{3.4 \text{ parts } (60 \text{ gal})}{8.1 \text{ parts}} = \textbf{25 gal of the 9.2\% solution}$$

$$\frac{4.7 \text{ parts } (60 \text{ gal})}{8.1 \text{ parts}} = \textbf{35 gal of the 1.1 \% solution}$$

To make 60 gal of the 4.5% solution, mix 25 gal of the 9.2% solution with 35 gal of the 1.1% solution.

2. The equation for calculating the percent K in potassium permanganate is:

$$\textbf{Percent K} = \frac{\textbf{Molecular weight of K (100\%)}}{\textbf{Molecular weight of KMnO}_4}$$

First, determine the molecular weight of each of the elements in the compound:

Element	Number of atoms		Atomic weight		Molecular weight
K	1	×	39.102	=	39.102
Mn	1	×	54.938	=	54.938
O	4	×	15.9994	=	63.9976
			Molecular weight of $KMnO_4$	=	158.0376

The molecular weight of $KMnO_4$ is 158.0376.

Substituting:

$$\text{Percent K} = \frac{39.102\ (100\%)}{158.0376} = \mathbf{24.742\%\ K}$$

3. Flow in the 10.0-in. pipe equals flow in the 6.0-in. pipe because the flow must remain constant.

$$\mathbf{Q_1 = Q_2}$$

Write the equation as above.

$$\mathbf{(Area\ 1)(Velocity\ 1) = (Area\ 2)(Velocity\ 2)}$$

First, find the diameter for the 6.0-in. and 10.0-in. pipes.

$$\text{Diameter for 6.0-in.} = (6.0\ \text{in.})(1\ \text{ft}/12\ \text{in.}) = 0.5\ \text{ft}$$
$$\text{Diameter for 10.0-in.} = (10.0\ \text{in.})(1\ \text{ft}/12\ \text{in.}) = 0.833\ \text{ft}$$

Then, determine the area of each pipe.

$$\mathbf{Area = (0.785)(Diameter)^2}$$
$$\text{Area 1 (6.0-in.)} = (0.785)(0.5\ \text{ft})(0.5\ \text{ft}) = 0.196\ \text{ft}^2$$
$$\text{Area 2 (10.0-in.)} = (0.785)(0.833\ \text{ft})(0.833\ \text{ft}) = 0.545\ \text{ft}^2$$

Substitute areas calculated and known velocity in 10.0-in. pipe.

$$(0.196\ \text{ft}^2)(x,\ \text{ft/s}) = (0.545\ \text{ft}^2)(1.35\ \text{ft/s})$$

Solve for x.

$$x,\ \text{ft/s} = \frac{(0.545\ \text{ft}^2)(1.35\ \text{ft/s})}{(0.196\ \text{ft}^2)} = 3.75\ \text{ft/s, round to } \mathbf{3.8\ ft/s\ in\ the\ 6.0\text{-}in.\ pipe}$$

4. First, determine the weight in pounds of the 11.9% solution.

$$\text{Weight of 11.9\% solution} = \text{Specific gravity (8.34 lb/gal)}$$
$$\text{Weight of 11.9\% solution} = 1.17(8.34\ \text{lb/gal}) = 9.76\ \text{lb/gal}$$

The equation is:

$$\mathbf{Percent\ mixture\ strength} =$$
$$\frac{\textbf{Soln. 1, gal(lb/gal)(Avail. \%/100\%) + Soln. 2, gal(lb/gal)(Avail. \%/100\%)}}{\textbf{Soln. 1, gal(lb/gal) + Soln. 2, gal(lb/gal)}} \times \mathbf{100\%}$$

(continued)

$$\text{Percent mixture strength} = \frac{235\text{ gal}(10.21\text{ lb/gal})(14.6\%/100\%) + 1{,}500\text{ gal}(9.76\text{ lb/gal})(11.9\%/100\%)}{235\text{ gal}(10.21\text{ lb/gal}) + 1{,}500\text{ gal}(9.76\text{ lb/gal})} \times 100\%$$

$$\text{Percent mixture strength} = \frac{350.3\text{ lb} + 1{,}742.16\text{ lb}}{2{,}399.35\text{ lb} + 14{,}640\text{ lb}} \times 100\% = \frac{2{,}092.46\text{ lb}}{17{,}039.35\text{ lb}} \times 100\%$$

$$= \mathbf{12\%\ strength\ of\ new\ ZOP\ solution}$$

5. First, calculate the iron removal in parts per million.

$$(1.58\text{ ppm})(89.1\%/100\%) = 1.58\text{ ppm}(0.891) = 1.408\text{ ppm}$$

Do the same for the manganese.

$$(0.35\text{ ppm})(75.2\%/100\%) = 0.35\text{ ppm}(0.752) = 0.263\text{ ppm}$$

Determine the amount of water, in million gallons, produced for the year.

$$(38\text{ mgd})(365\text{ d/yr}) = 13{,}870\text{ mil gal/yr}$$

Next, using the "pounds" equation, solve for the number of pounds per year for iron and manganese.

$$\mathbf{Pounds\ per\ year = (mil\ gal/yr)(Dosage,\ mg/L)(8.34\ lb/gal)}$$

$$\text{Pounds per year, Fe removal} = (13{,}870\text{ mil gal/yr})(1.408\text{ mg/L})(8.34\text{ lb/gal})$$

$$= 162{,}872\text{ lb/yr, round to } \mathbf{160{,}000\ lb/yr\ of\ Fe\ removed}$$

$$\text{Pounds per year, Mn removal} = (13{,}870\text{ mil gal/yr})(0.263\text{ mg/L})(8.34\text{ lb/gal})$$

$$= 30{,}423\text{ lb/yr, round to } \mathbf{30{,}000\ lb/yr\ of\ Mn\ removed}$$

6. First, determine the number of milliliters per minute for the polymer.

$$\frac{171\text{ mL}}{4\text{ min}} = 42.75\text{ mL/min}$$

Next, calculate the pounds per gallon for the polymer.

$$\text{Pounds per gallon} = (\text{Specific gravity})(8.34\text{ lb/gal})$$

$$\text{Pounds per gallon, polymer} = (1.37)(8.34\text{ lb/gal}) = 11.426\text{ lb/gal}$$

Then, find the number of million gallons per day.

$$(7{,}250\text{ gpm})(1440\text{ min/d})(1\text{ M}/1{,}000{,}000) = 10.44\text{ mgd}$$

The dosage equation is:

$$\textbf{Milligrams per liter} = \frac{\textbf{(mL/min)(1,440 min/d)(lb/gal, polymer)}}{\textbf{(3,785 mL/gal)(mgd)(8.34 lb/gal)}}$$

$$\text{Milligrams per liter, polymer} = \frac{(42.75\text{ mL/min})(1{,}440\text{ min/d})(11.426\text{ lb/gal})}{(3{,}785\text{ mL/gal})(10.44\text{ mgd})(8.34\text{ lb/gal})}$$

$$= \textbf{2.13 mg/L, polymer}$$

7. First, find the amount of water produced during the 5.5-hr interval.

Water produced in 5.5-hr interval = (23.9 mgd)(1,000,000 gal/M)(5.5 hr)/24 hr
= 5,477,083 gal

Next, find the number of gallons contained in the 4.27-ft rise in water level.

$$\textbf{Volume, tank} = \textbf{(0.785)(Diameter)}^2\textbf{(Height)}$$

Volume of 4.27 ft in 225 ft diameter tank = (0.785)(225 ft)(225 ft)(4.27 ft)(7.48 gal/ft^3)
= 1,269,300 gal

Production minus the rise in level is the amount the discharge pumps had to send into the distribution system.

Total gallons discharge pumps moved in 5.5 hr = 5,477,083 gal – 1,269,300 gal
= 4,207,783 gal

Next, find the number of minutes in 5.5 hr.

(5.5 hr)(60 min/hr) = 330 min

Last, divide the number of gallons the discharge pumps moved by the time in minutes.

Discharge pumps, gpm = 4,207,783 gal/330 min

= 12,751 gpm, round to **13,000 gpm**

8. First, convert kilograins to grains.

(28.1 kilograins/ft^3)(1,000 grains/kilograins) = 28,100 grains/ft^3

Next, find the capacity of the unit in grains.

(575 ft^3)(28,100 grains/ft^3) = 16,157,500 grains

(continued)

The equation is:

$$\textbf{Water treatment capacity, gal} = \frac{\textbf{(Exchange capacity in grains)}}{\textbf{Hardness, gpg}}$$

$$\text{Water treatment capacity, gal} = \frac{(16{,}157{,}500 \text{ grains})}{19.6 \text{ gpg}}$$

$$= 824{,}362 \text{ gal, round to } \mathbf{824{,}000\ gal}$$

9. First, find the pounds per gallon for alum.

$$\text{Pounds per gallon, alum} = (\text{Specific gravity})(8.34 \text{ lb/gal})$$

$$\text{Pounds per gallon, alum} = (1.32 \text{ sp gr})(8.34 \text{ lb/gal}) = 11.01 \text{ lb/gal}$$

The dosage equation is:

$$\textbf{Dosage, mg/L} = \frac{\textbf{(mL/min)(1,440 min/d)(Alum, lb/gal)(\% purity)}}{\textbf{(mgd)(8.34 lb/gal)(3,785 mL/gal)}}$$

Rearrange to solve for the unknown (milliliters per minute).

$$\text{Alum, mL/min} = \frac{(\text{Dosage, mg/L})(\text{mgd})(8.34 \text{ lb/gal})(3{,}785 \text{ mL/gal})}{(\text{Alum, lb/gal})(\% \text{ Purity})(1{,}440 \text{ min/d})}$$

$$\text{Alum, mL/min} = \frac{(15.6 \text{ mg/L})(17.5 \text{ mgd})(8.34 \text{ lb/gal})(3{,}785 \text{ mL/gal})}{(11.01 \text{ lb/gal})(48.0\%/100\% \text{ purity})(1{,}440 \text{ min/d})}$$

$$\text{Alum, mL/min} = 1{,}132.4 \text{ mL/min, round to } \mathbf{1{,}130\ mL/min\ of\ alum}$$

10. First, convert 1,985 gpd of the hypochlorite solution rate of flow to million gallons per day.

$$\frac{(1{,}985 \text{ gpd})(1 \text{ mil gal})}{1{,}000{,}000 \text{ gal}} = 0.001985 \text{ mgd}$$

Then, use the equal dosage equations:

$$(0.001985 \text{ mgd})(x, \text{ mg/L})(9.25 \text{ lb/gal}) = (46 \text{ mgd})(3.0 \text{ mg/L})(8.34 \text{ lb/gal})$$

$$\text{Milligrams per liter, hypochlorite} = \frac{(46 \text{ mgd})(3.0 \text{ mg/L})(8.34 \text{ lb/gal})}{(0.001985 \text{ mgd})(9.25 \text{ lb/gal})} = 62{,}682 \text{ mg/L}$$

Last, convert milligrams per liter to percent.

$$\frac{(62{,}682 \text{ mg/L}) \quad (1\%)}{10{,}000 \text{ mg/L}} = \mathbf{6.3\%\ hypochlorite}$$

Appendixes

APPENDIX

A

COMMON CONVERSION FACTORS

AREA

1 acre = 43,560 square feet (ft^2)
1 acre-ft = 43,560 cubic feet (ft^3)
2.4711 acre = 1 hectare (ha)
1 ha = 0.4047 acre
1 hectare = 10,000 square meters (m^2)
1 square mile (mi^2) = 640 acres

CONCENTRATION

1% solution = 1 part in 100 parts
1% solution = 10,000 ppm
1 ppm = 1 mg/L
1 gpg = 17.12 ppm

DENSITY

Water has a density of 1 g/mL or 8.34 lb/gal or 62.4 lb/ft^3

FLOW

1 miner's inch = 1.5 ft^3/min
1 ft^3/s = 448.8 gal/min (gpm)
1 ft^3/s = 0.6463 million gallons per day (mgd)
1 mgd = 1.547 ft^3/s

LENGTH

1 inch (in.) = 2.54 centimeter (cm)
100 cm = 1 m
1 m = 39.37 in.
1 m = 3.281 feet (ft)
1 yard (yd) = 0.9144 meter (m)
1,000 m = 1 kilometer (km)
1 km = 1.609 miles (mi)

POWER

1 horsepower (hp) = 0.746 kilowatts (kW)
1 kW = 1.341 hp

PRESSURE

1 lb/square inch (psi) = 2.307 ft of water
1 ft of water = 0.4335 psi
1 atmosphere (atm) = 14.7 psi
1 atm = 29.92 in. of mercury (Hg)
1 atm = 33.90 ft of water
1 atm = 760 millimeters (mm) of mercury

TEMPERATURE

Degrees Fahrenheit (°F) = $\frac{9}{5}$(C° + 32)
Degrees Celsius (°C) = $\frac{5}{9}$(°F – 32)

VOLUME

2 pints = 1 quart
8 pints = 1 gallon (gal)
4 quarts = 1 gal
1 quart = 32 fluid ounces (oz)
1 gal = 128 fluid oz
1 gal = 3.785 liters (L)
1 L = 1.0567 quarts
1 L = 1,000 milliliters (mL)
3,785 mL = 1 gal
1,000 L = 1 cubic meter (m^3)
1 ft^3 = 7.48 gal
1 million gallons (mil gal) = 3.07 acre-ft
1 acre-ft = 325,829 gal

WEIGHT

1 gram (g) = 1,000 milligrams (mg)
1,000 g = 1 kilogram (kg)
1 pound (lb) = 454 g
1 lb = 7,000 grains (gr)
1 kg = 2.205 lb
2,000 lb = 1 ton
1 mg/L = 1 part per million (ppm)
1 grain per gal (gpg) = 17.1 ppm

APPENDIX B

COMMON FORMULAS

COMMON CONVERSIONS

Acre-feet to cubic feet:

$$(\text{Number of acre-ft})(43{,}560\ \text{ft}^3/\text{acre-ft}) = \text{Number of ft}^3$$

Acre-feet to gallons:

$$\text{Number of gal} = (\text{Number of acre-feet})(43{,}560\ \text{ft}^3/\text{acre-ft})(7.48\ \text{gal/ft}^3)$$

Cubic feet per second to gallons per minute:

$$\text{Number of gpm} = (\text{Number of ft}^3/\text{s})(60\ \text{s/min})(7.48\ \text{gal/ft}^3)$$

Cubic feet per second to million gallons per day:

$$\frac{(\text{Number of ft}^3)(60\ \text{s})(1{,}440\ \text{min})\,(7.48\ \text{gal})\ \ (1\ \text{mil gal})}{\text{sec}\qquad\text{min}\qquad\text{day}\qquad\text{ft}^3\qquad 1{,}000{,}000\ \text{gal}} = \text{Number of mgd}$$

Gallons per minute to cubic feet per second:

$$\text{Number of ft}^3/\text{s} = \frac{(\text{Number of gpm})}{(60\ \text{s/min})(7.48\ \text{gal/ft}^3)}$$

Gallons per minute to liters per second:

$$\text{Flow, L/s} = \frac{(\text{Flow, gpm})(3.785\ \text{L/gal})}{60\ \text{s/min}}$$

Gallons to cubic feet:

$$\frac{(\text{Number of gal})(1\ \text{ft}^3)}{7.48\ \text{gal}} = \text{Number of ft}^3$$

Gallons to liters or liters to gallons:

$$\text{Number of liters} = (\text{Number of gal})(3.785\ \text{L/gal})$$

$$\text{Number of gal} = (\text{Number of liters})(1\ \text{gal}/3.785\ \text{L})$$

Gallons to pounds:

$$\frac{(\text{Number of gal})(8.34\ \text{lb})}{\text{gal}} = \text{Number of lb}$$

Grams to pounds:

$$\text{Number of lb} = (\text{Number of grams})(1\ \text{lb}/454\ \text{g})$$

Liters per second to gallons per minute:

$$\text{Flow, gpm} = \frac{(\text{Flow, L/s})(60\ \text{s/min})}{3.785\ \text{L/gal}}$$

Million gallons per day to cubic feet per second:

$$\frac{(\text{Number of mgd})(1{,}000{,}000\ \text{gal})\ (1\ \text{ft}^3)\quad (1\ \text{d})\quad (1\ \text{min})}{(1\ \text{mil gal})\ (7.48\ \text{gal})(1{,}440\ \text{min})\ (60\ \text{s})} = \text{Number ft}^3/\text{s}$$

Parts per million to percent:

$$\%\ \text{solution} = \frac{(\text{Known ppm})(1\%)}{10{,}000\ \text{ppm}}$$

ALKALINITY (IN TABLE FORM)

Alkalinity, mg/L as $CaCO_3$

Results of Titration	Bicarbonate alkalinity	Carbonate alkalinity	Hydroxide alkalinity
P = 0	T	0	0
P is less than ½T	T–2P	2P	0
P = ½T	0	2P	0
P is greater than ½T	0	2T–2P	2P–T
P = T	0	0	0

Where P = phenolphthalein alkalinity and T = total alkalinity.

AREA

Area of a rectangle = (Length)(Width)

Area of a circle (tank) = $(0.785)(\text{Diameter})^2$ or πr^2

AVERAGE

$$\text{Average} = \frac{\text{Sum of measurements}}{\text{Number of measurements}}$$

BACKWASH

$$\text{Backwash rate, gpm/ft}^2 = \frac{\text{Backwash pumping rate, gpm}}{\text{Filter area, ft}^2}$$

BLENDING

$$Q_1C_1 + Q_2C_2 = Q_3C_3$$

Where: Q_1 = Flow from the well with highest concentration of species in question

C_1 = Highest concentration of species in well with the highest concentrations

Q_2 = Lowest flow from well with low levels of species in question

C_2 = Highest concentration of species from well with lowest levels

$Q_3 = Q_1 + Q_2$

C_3 = Highest permissible level of species in question

CHEMISTRY

$$\text{Percent of an element in a compound} = \frac{(\text{Molecular weight of the element})(100\%)}{\text{Molecular weight of the compound}}$$

(continued)

$$\text{Normality (N)} = \frac{\text{Number of gram-equivalents of solute}}{\text{Number of L, solution}}$$

$$\text{Dosage, mg/L} = \frac{\text{(Stock, mL)(1,000 mg/g)(Concentration, g/L)}}{\text{Sample size, mL}}$$

CORROSION

Langelier index = pH – pH_s

$$\text{mils penetration per year} = \frac{\text{(Weight loss)(534)(1,000)}}{\text{(Coupon area, in.}^2\text{)(Time, hr)(Metal density, g/cm}^3\text{)}}$$

DENSITY

Density = Mass/Volume

DOSAGE

Chlorine dose = Chlorine demand + Chlorine residual

Chemical feed, lb/d = (Flow, mgd)(Dosage, mg/L)(8.34 lb/gal)

or rearranging to solve for dosage

$$\text{Dosage, mg/L} = \frac{\text{lb/d}}{\text{(mgd)(8.34 lb/gal)}}$$

$$\text{Pounds per day} = \frac{\text{(mgd)(Dosage, mg/L)(8.34 lb/gal)}}{\text{(Percent purity)}}$$

Used when the purity of a substance or solution is <100%.

$$\text{(mgd)}(x\text{, mg/L)(8.34 lb/gal)} = \text{(mgd)(Dosage, mg/L)(8.34 lb/gal)}$$

$$\text{Dosage, mg/L} = \frac{\text{(mL/min)(1,440 min/d)(Alum, lb/gal)(Percent purity)}}{\text{(mgd)(8.34 lb/gal)(3,785 mL/gal)}}$$

$$KMnO_4\text{, mg/L} = 0.2\text{(Iron, mg/L)} + 2.0\text{(Manganese, mg/L)}$$

FILTRATION

$$\text{Filtration rate} = \frac{\text{Flow rate, gpm}}{\text{Filter surface area, ft}^2}$$

$$\text{Unit filter run volume, gal/ft}^2 = \frac{\text{Total gallons filtered}}{\text{Filter surface area, ft}^2}$$

FLOW RATE

Flow = Volume/Time

Q (Flow) = (Area)(Velocity)

Example:

Q, Flow in $\text{ft}^3\text{/s}$ = (Area, ft^2)(Velocity, ft/s)

$$\text{Weir overflow rate} = \frac{\text{Flow, gpd}}{\text{Weir length, ft}}$$

Flow in a pipe that changes size:

$$(\text{Area 1, ft}^2)(\text{Velocity 1, ft/s}) = (\text{Area 2, ft}^2)(\text{Velocity 2, ft/s})$$

$$\text{Weir overflow rate} = \frac{\text{Flow, gpd}}{\text{Weir length, ft}}$$

Flow through a trapezoidal channel:

$$\text{Flow, ft}^3\text{/s} = \frac{(\text{Bottom width} + \text{Water surface width})}{2}(\text{Depth, ft})(\text{Velocity, ft/s})$$

Gallons per minute to million gallons per day (MGD)

$$\text{MGD} = \frac{(\text{Pumping rate, gpm})(1{,}440\text{ min/d})}{1{,}000{,}000\text{/Million}}$$

GALLONS PER CAPITA PER DAY

$$\text{Gallons/Capita/Day} = \frac{\text{Volume, gpd}}{\text{Population served/day}}$$

HARDNESS

Calcium (Ca) hardness as mg/L $CaCO_3$ = 2.5(Ca content in mg/L)

Calcium hardness as mg/L $CaCO_3$ = 4.1(Magnesium [Mg] content in mg/L)

Total hardness, mg/L as $CaCO_3$ = Ca hardness, mg/L as $CaCO_3$ + Mg hardness, mg/L as $CaCO_3$

HYDRATED LIME

$$\text{Hydrated lime feed, mg/L} = \frac{(A + B + C + D)(1.15)}{\text{Lime purity}}$$

Where A is CO_2 in source water:

$$A = (\text{mg/L as } CO_2)(74.1/44)$$

where 74.1 = Molecular weight (MW) of $Ca(OH)_2$ and 44 MW of CO_2

Where B is bicarbonate (total) alkalinity removed in softening:

$$B = (\text{mg/L as } CaCO_3 \text{ removed})(74.1/100.1)$$

where 100.1 = MW of $CaCO_3$

Where C is hydroxide alkalinity in softened effluent:

$$C = (\text{mg/L as } CaCO_3 \text{ removed})(74.1/100.1)$$

Where D is magnesium removed in softening:

$$D = (\text{mg/L as } Mg^2\ )(74.1/24.3)$$

where 24.3 = MW of Mg^2

KILOWATT

kW = (Number of hp)(0.746 kW/hp)

LIME DOSAGE PROBLEMS

$$\frac{0.45 \text{ mg/L alkalinity}}{1 \text{ mg/L alum}} = \frac{x \text{ mg/L Alkalinity}}{\text{Number of mg/L alum required}}$$

Total alkalinity required, mg/L = Alkalinity reacting with alum, mg/L + Residual alkalinity for precipitation, mg/L

Alkalinity needed for raw water, mg/L = Total alkalinity required, mg/L – Alkalinity in raw water, mg/L

$$\frac{0.35 \text{ mg/L lime}}{0.45 \text{ mg/L alkalinity}} = \frac{\text{Number of mg/L lime}}{\text{Alkalinity needed, mg/L}}$$

LOG REMOVAL AND CT

CT calculated = (Chlorine concentration, mg/L)(Time, min)

$$\text{Inactivation ratio (IR)} = \frac{\text{Calculated CT value}}{\text{CT table value}}$$

$$\text{Log removal} = \text{Log}_{10}\,\frac{(1.0 - \text{Percent removal})(-)}{100\%}$$

MILLIAMP SIGNAL

$$\frac{\text{mA maximum}}{\text{Maximum possible read}} = \frac{\text{Unknown number of mA}}{\text{Current read}}$$

MIXTURE

Percent mixture strength =

$$\frac{\text{Solution 1 gal (Available \%/100\%)} + \text{Solution 2 gal (Available \%/100\%)}}{\text{gal of solution 1} + \text{gal of solution 2}} \times 100\%$$

Percent mixture =

$$\frac{\text{Soln. 1, gal(lb/gal of soln.)(Avail. \%/100\%)} + \text{Soln. 2, gal(lb/gal of soln.)(Avail. \%/100\%)}}{\text{Soln. 1, gal(lb/gal of soln. 1)} + \text{Soln. 2, gal(lb/gal of soln. 2)}}$$

$$\times 100\%$$

OPERATING TIME

$$\text{Operating time} = \frac{\text{Treated water}}{\text{Flow rate}}$$

PERCENT CALCULATION

$$\text{Percent} = \frac{(\text{Low number})(100\%)}{\text{High number}}$$

PERCENT RECOVERY FOR REVERSE OSMOSIS

$$\text{Percent recovery} = \frac{(\text{Product flow})(100\%)}{\text{Feed flow}}$$

PERCENT REMOVAL

$$\text{Percent ntu removal} = \frac{\text{Influent ntu} - \text{Effluent ntu}}{\text{Influent ntu}} \times 100\%, \text{ or}$$

$$\frac{\text{In} - \text{Out}}{\text{In}} \times 100\%$$

PERIMETER

Circumference = (Diameter) = (3.14)(Diameter)

PRESSURE

$$\text{psi} = \frac{\text{Depth, ft}}{2.31 \text{ ft/psi}}$$

$$\text{psi} = (\text{Depth, ft})(0.433 \text{ psi/ft})$$

$$\text{Pressure lb/ft}^2 = \frac{\text{Force, lb}}{\text{Area, ft}^2}$$

PUMPING

Total head, ft = Total static head, ft + Head loss, ft

$$\text{mhp} = \frac{(\text{whp})}{(\text{Motor efficiency})(\text{Pump efficiency})}$$

where mhp = motor horsepower and whp = water horsepower

$$\text{mhp} = \frac{(\text{Flow, gpm})(\text{TH, ft})}{(3{,}960)(\text{Motor efficiency})(\text{Pump efficiency})}$$

where TH = Total head

$$\text{Brake hp} = (\text{hp})(\text{Motor efficiency})$$

where hp = horsepower

Water hp = (mhp)(Motor efficiency)(Pump efficiency)

Water hp = (bhp)(Pump efficiency)

Motor hp = bhp/Motor efficiency

Brake hp = whp/Pump efficiency

RATIO

$$\frac{A}{B} = \frac{C}{D} \text{ or } A{:}B = C{:}D$$

SEDIMENTATION

$$\text{Detention time, hr} = \frac{\text{Volume, gal}}{\text{Flow rate, gph}}$$

$$\text{Surface loading rate} = \frac{\text{Gallons per day}}{\text{Number of ft}^2}$$

$$\text{Percent settled sludge} = \frac{(\text{Settled sludge, mL})(100\%)}{\text{Total sample volume, mL}}$$

SOFTENING

$$\text{Hardness, mg/L} = \frac{(\text{Hardness, gpg})(17.12 \text{ mg/L})}{1 \text{ gpg}}$$

where gpg = grains per gallon

$$\text{Water treatment capacity, gal} = \frac{\text{Exchange capacity, grains}}{\text{Hardness, grains per gallon}}$$

SOLUTION

$(\text{Concentration}_1)(\text{Volume}_1) + (\text{Concentration}_2)(\text{Volume}_2) = (\text{Concentration}_3)(\text{Volume}_3)$ or

$$C_1V_1 + C_2V_2 = C_3V_3$$

Solution or dilution triangle:

Conc. 1 Conc. 2 – Conc. desired = No. of parts conc.
Conc. desired
Conc. 2 Conc. 1 – Conc. desired = No. of parts
Total no. of parts

$$\frac{(\text{Number of gal})(\text{Number of parts, conc. 1})}{\text{Total number of parts}} = \text{Number of gallons of conc. 1 required}$$

$$\frac{(\text{Number of gal})(\text{Number of parts, conc. 2})}{\text{Total number of parts}} = \text{Number of gallons of conc. 2 required}$$

$$\text{Percent HTH solution} = \frac{(\text{lb HTH})(100\%)}{(\text{Number of gal})(8.34 \text{ lb/gal})}$$

SPECIFIC GRAVITY

Specific gravity (sp gr) = Density of substance/Density of water

SPECIFIC YIELD

$$\text{Specific yield, gpm/ft} = \frac{\text{Well yield, gpm}}{\text{Drawdown, ft}}$$

or

$$\text{Drawdown, ft} = \frac{\text{Well yield, gpm}}{\text{Specific yield, gpm/ft}}$$

or

$$\text{Well yield, gpm} = (\text{Specific yield, gpm/ft})(\text{Drawdown, ft})$$

TEMPERATURE

$°C = \frac{5}{9}(°F - 32)$

$°F = (\frac{9}{5})°C + 32$

TOTAL FORCE

Total force, lb = (Area, in.^{2})(psig)

VOLUME

Volume of a basin = (Length)(Width)(Depth)

Volume of a basin in gallons = (Length)(Width)(Depth)(7.48 gal/ft^3)

Volume of a cone = $\frac{1}{3}\pi r^2$(Height or Depth)

Volume of a circular tank = πr^2(Height) or Volume of a pipe = πr^2(Length)

Volume of a cylindrical tank = (0.785)(Diameter)^2(Height)

$$\text{Volume of a trough} = \frac{(b_1 + b_2)}{2}(\text{Depth of water})(\text{Length})(7.48\ \text{gal/ft}^3)$$

WATER SUPPLY

$$\text{Supply, hr} = \frac{\text{Storage volume, gal}}{(\text{Pumping rate, gpm})(60 \text{ min/hr})}$$

WELL DRAWDOWN

Drawdown, ft = Pumping water level, ft – Static water level, ft

Pumping water level, ft = Drawdown, ft + Static water level

Static water level, ft = Pumping water level, ft – Drawdown, ft

APPENDIX

C

CHEMISTRY TABLES

TABLE C-1 International Atomic Weights (Based on Carbon-12)

Element	Symbol	Atomic Number	Atomic Weight*	Element	Symbol	Atomic Number	Atomic Weight*
Actinium	Ac	89	227.0278	Erbium	Er	68	167.26
Aluminum	Al	13	26.98154	Europium	Eu	63	151.96
Americium	Am	95	(243)	Fermium	Fm	100	(257)
Antimony	Sb	51	121.75	Fluorine	F	9	18.998403
Argon	Ar	18	39.948	Francium	Fr	87	(223)
Arsenic	As	33	74.9216	Gadolinium	Gd	64	157.25
Astatine	At	85	(210)	Gallium	Ga	31	69.72
Barium	Ba	56	137.33	Germanium	Ge	32	72.59
Berkelium	Bk	97	(247)	Gold	Au	79	196.9665
Beryllium	Be	4	9.01218	Hafnium	Hf	72	178.49
Bismuth	Bi	83	208.9804	Hassium	Hs	108	(265.0)
Bohrium	Bh	107	(262.0)	Helium	He	2	4.00260
Boron	B	5	10.81	Holmium	Ho	67	164.9304
Bromine	Br	35	79.904	Hydrogen	H	1	1.00794
Cadmium	Cd	48	112.41	Indium	In	49	114.82
Calcium	Ca	20	40.08	Iodine	I	53	126.9045
Californium	Cf	98	(251)	Iridium	Ir	77	192.22
Carbon	C	6	12.011	Iron	Fe	26	55.847
Cerium	Ce	58	140.12	Krypton	Kr	36	83.80
Cesium	Cs	55	132.9054	Lanthanum	La	57	138.9055
Chlorine	Cl	17	35.453	Lawrencium	Lw	103	(260)
Chromium	Cr	24	51.996	Lead	Pb	82	207.2
Cobalt	Co	27	58.9332	Lithium	Li	3	6.941
Copper	Cu	29	63.546	Lutetium	Lu	71	174.967
Curium	Cm	96	(247)	Magnesium	Mg	12	24.305
Dubnium	Db	105	(262.0)	Manganese	Mn	25	54.9380
Dysprosium	Dy	66	162.50	Mendelevium	Mv	101	(258)
Einsteinium	Es	99	(252)	Meitnerium	Mt	109	(266.0)

(continued)

TABLE C-1 International Atomic Weights (Based on Carbon-12) (continued)

Element	Symbol	Atomic Number	Atomic Weight*	Element	Symbol	Atomic Number	Atomic Weight*
Mercury	Hg	80	200.59	Scandium	Sc	21	44.9559
Molybdenum	Mo	42	95.94	Seaborgium	Sg	106	(263.0)
Neodymium	Nd	60	144.24	Selenium	Se	34	78.96
Neon	Ne	10	20.179	Silicon	Si	14	28.0855
Neptunium	Np	93	237.0482	Silver	Ag	47	107.8682
Nickel	Ni	28	58.69	Sodium	Na	11	22.98977
Niobium	Nb	41	92.9064	Strontium	Sr	38	87.62
Nitrogen	N	7	14.0067	Sulfur	S	16	32.06
Nobelium	No	102	(259)	Tantalum	Ta	73	180.9479
Osmium	Os	76	190.2	Technetium	Tc	43	(98)
Oxygen	O	8	15.9994	Tellurium	Te	52	127.60
Palladium	Pd	46	106.42	Terbium	Tb	65	158.9254
Phosphorus	P	15	30.97376	Thallium	Tl	81	204.383
Platinum	Pt	78	195.08	Thorium	Th	90	232.0381
Plutonium	Pu	94	(244)	Thulium	Tm	69	168.9342
Polonium	Po	84	(209)	Tin	Sn	50	118.69
Potassium	K	19	39.0983	Titanium	Ti	22	47.88
Praseodymium	Pr	59	140.9077	Tungsten	W	74	183.85
Promethium	Pm	61	(145)	Ununnilium	Uun	110	(269.0)
Protactinium	Pa	91	231.0359	Unununium	Uuu	111	(272.0)
Radium	Ra	88	226.0254	Unumbium	Uub	112	(277.0)
Radon	Rn	86	(222)	Uranium	U	92	238.0289
Rhenium	Re	75	186.207	Vanadium	V	23	50.9415
Rhodium	Rh	45	102.9055	Ytterbium	Yb	70	173.04
Rubidium	Rb	37	85.4678	Yttrium	Y	39	88.9059
Ruthenium	Ru	44	101.07	Xenon	Xe	54	131.29
Rutherfordium	Rf	104	(261.0)	Zinc	Zn	30	65.38
Samarium	Sm	62	150.36	Zirconium	Zr	40	91

Source: US Government Printing Office.
*The numbers in parenthesis indicate mass number of most stable known isotope.

TABLE C-2 Common Chemicals Used and Encountered in the Water Industry

Chemical	Chemical Formula	Use	Miscellaneous
Alum	$Al_2(SO_4)_3 \cdot 14(H_2O)$	Coagulant	
Ammonia	NH_3	Check for chlorine leaks	Make chloramine
Bicarbonate	HCO_3		Carbonate hardness
Calcium carbonate	$CaCO_3$	Primary hardness chemical	
Calcium hypochlorite	$Ca(OCl)_2$	Disinfectant	
Carbon dioxide	CO_2	Recarbonation	
Caustic soda	NaOH	Adjust pH	
Chlorine	Cl_2	Disinfectant	
Chlorine dioxide	ClO_2	Disinfectant	
Copper sulfate	$CuSO_4$	Algae control	
Dichloramine	$NHCl_2$	Disinfectant	Formed when ammonia is added to water containing Cl
Ferric chloride	$FeCl_3$	Coagulant	
Ferric sulfate	$Fe_2(SO_4)_3$	Coagulant	
Ferrous sulfate	$Fe_2(SO_4)_3 \cdot 7(H_2O)$	Coagulant	
Fluorosilicic acid*	H_2SiF_6	Fluoridation	
Hydrated lime	$Ca(OH)_2$	Increase pH and alkalinity	
Hydrochloric acid	HCl	Laboratory, cleaning	
Hydroxide ion	OH		Naturally found in water
Hypochlorite ion	OCl	Disinfectant	
Magnesium hydroxide	$Mg(OH)_2$		Formed in the lime–soda softening process
Monochloramine	NH_2Cl	Disinfectant	Formed when ammonia is added to water containing Cl
Nitrate	NO_3		Fertilizer, sewage, natural deposits that are eroded
Quicklime	CaO	Water stabilization, increase pH and alkalinity	Water softening
Ozone	O_3	Disinfectant	
Potassium permanganate	$KMnO_4$	Control taste-and-odor substances	Oxidize Fe and Mn
Sulfuric acid	H_2SO_4	Decrease pH and alkalinity	Water stabilization
Soda ash	Na_2CO_3	Adjust pH	
Sodium aluminate	$Na_2Al_2O_4$	Coagulant	
Sodium bicarbonate	$NaHCO_3$	Increase pH and alkalinity	Water stabilization
Sodium fluoride	NaF	Fluoridation	
Sodium hexametaphosphate	$(NaPO_3)_n \cdot Na_2O$†	Sequestering agent	
Sodium hydroxide	NaOH	Adjust pH	
Sodium silicate	$Na_2O \cdot (SiO_2)_x$	Coagulant	
Sodium fluorosilicate‡	Na_2SiF_6	Fluoridation	
Trichloramine	NCl_3		Formed when ammonia is added to water containing Cl
Zinc orthophosphate	$Zn_3(PO_4)_2$	Forms protective coating	

* Formerly known as hydrofluosilicic acid or "silly acid."
† Typically n = 14.
‡ Formerly known as sodium silicofluoride.

TABLE C-3 Formula Weights

Compound	Weight, grams	Compound	Weight, grams
AgCl	143.32	$KHC_8H_4O_4$ (phthalate)	204.224
Ag_2CrO_4	331.73	$KH(IO_3)_2$	389.912
$Al_2(SO_4)_3 \cdot 14(H_2O)$	594.35	K_2HPO_4	174.176
$BaSO_4$	233.39	KH_2PO_4	136.086
$CaCO_3$	100.089	$KHSO_4$	136.16
CaC_2O_4	128.100	KI	166.003
CaF_2	78.077	KIO_3	214.001
CaO	56.079	KIO_4	230.000
$Ca(OCl)_2$	142.985	$KMnO_4$	158.034
$Ca(OH)_2$	74.095	KNO_3	101.101
CO_2	44.010	$Mg(OH)_2$	58.320
Cl_2	70.906	$MgSO_4$	120.36
ClO_2	67.452	MnO_2	86.937
Cr_2O_3	151.990	Mn_2O_3	157.874
CuO	79.545	Mn_3O_4	228.812
$CuSO_4$	159.60	NaBr	102.894
$NHCl_2$	85.921	NaCl	58.443
$FeCl_3$	162.206	NaCN	49.007
FeO	71.846	Na_2CO_3	105.989
Fe_2O_3	159.692	$Na_2Al_2O_4$	163.940
Fe_3O_4	231.539	$NaHCO_3$	84.007
$Fe_2(SO_4)_3$	399.87	NaF	41.988
$Fe_2(SO_4)_3 \cdot 7(H_2O)$	525.97	NaOH	39.997
HBr	80.912	$Na_2O \cdot (SiO_2)_x$	Variable
$HC_2H_3O_2$ (acetic acid)	60.052	Na_2SiF_6	188.056
HCO_3	61.017	NCl_3	120.366
HCl	36.461	NH_3	17.030
$HClO_4$	100.458	NH_2Cl	51.476
HNO_3	63.013	NH_4Cl	53.491
H_2O	18.015	NH_4NO_3	80.043
H_2O_2	34.015	$(NH_4)_2SO_4$	132.13
H_3PO_4	97.995	NO_3	62.005
H_2S	34.08	O_3 (ozone)	47.998
H_2SO_3	82.07	OH	17.007
H_2SO_4	98.07	OCl	51.452
H_2SiF_6	144.092	$PbCrO_4$	323.2
HgO	216.59	$Pb(NO_3)_2$	331.2
Hg_2Cl_2	472.09	PbO	223.2
$HgCl_2$	271.50	PbO_2	239.2
KBr	119.002	$PbSO_4$	303.3
$KBrO_3$	167.000	P_2O_5	141.944
KCl	74.551	Sb_2S_3	339.68
$KClO_3$	122.550	SiO_2	60.084
KCN	65.116	$SnCl_2$	189.596
K_2CrO_4	194.190	SnO_2	150.689
$K_2Cr_2O_7$	294.184	SO_2	64.06
$K_3Fe(CN)_6$	329.248	SO_3	80.06
$K_4Fe(CN)_6$	368.346	$Zn_3(PO_4)_2$	304.703

Based on US Government Printing Office table of atomic weights (see Table C-1).

APPENDIX D

MAXIMUM CONTAMINANT LEVELS

TABLE D-1 Maximum Contaminant Levels for Inorganic Chemicals

Chemical	MCL, mg/L	MCL, mg/L*	MCL, mg/L*
Aluminum	1.		
Antimony	0.006		
Arsenic	0.05		
Asbestos	7 million fibers per liter		
Barium	1.		
Beryllium	0.004		
Cadmium	0.005		
Chromium	0.05		
Cyanide	0.2		
Fluoride	2.		
Mercury	0.002		
Nickel	0.1		
Nitrate (as NO_3)	45.		
Nitrate + Nitrite (sum as nitrogen)	10.		
Nitrite (as nitrogen)	1.		
Selenium	0.05		
Thallium	0.002		

*Extra columns are for any future changes in the MCLs.

TABLE D-2 Maximum Contaminant Levels for Volatile Organic Chemicals (VOCs)

Volatile Organic Chemical	MCL, mg/L	MCL, mg/L*	MCL, mg/L*
Benzene	0.001		
Carbon tetrachloride	0.0005		
1,2-Dichlorobenzene	0.6		
1,4-Dichlorobenzene	0.005		
1,1-Dichloroethane	0.005		
1,2-Dichloroethane	0.0005		
1,1-Dichloroethylene	0.006		
cis-1,2-Dichloroethylene	0.006		
trans-1,2-Dichloroethylene	0.01		
Dichloromethane	0.005		
1,2-Dichloropropane	0.005		
1,3-Dichloropropene	0.0005		
Ethylbenzene	0.7		
Monochlorobenzene	0.07		
Styrene	0.1		
1,1,2,2-Tetrachloroethane	0.001		
Tetrachloroethylene	0.005		
Toluene	0.15		
1,2,4-Trichlorobenzene	0.07		
1,1,1-Trichloroethane	0.200		
1,1,2-Trichloroethane	0.005		
Trichloroethylene	0.005		
Trichlorofluoromethane	0.15		
1,1,2-Trichloro-1,2,2-Trifluoroethane	1.2		
Vinyl chloride	0.0005		
Xylenes	1.750†		

*Extra columns are for any future changes in the MCLs.

†The MCL is for one xylene isomer or the sum of all the xylene isomers.

TABLE D-3 Maximum Contaminant Levels for Nonvolatile Synthetic Organic Chemicals (SOCs)

Nonvolatile	MCL, mg/L	MCL, mg/L*	MCL, mg/L*
Alachlor	0.002		
Atrazine	0.003		
Bentazon	0.018		
Benzo(a)pyrene	0.000		
Carbofuran	0.018		
Chlordane	0.0001		
2,4-D	0.07		
Dalapon	0.2		
Dibromochloropropane (DBCP)	0.0002		
Di(2-ethylhexyl)adipate	0.4		
Di(2-ethylhexyl)phthalate	0.004		
Dinoseb	0.007		
Diquat	0.02		
Endothall	0.1		
Endrin	0.002		
Ethylene dibromide (EDB)	0.00005		
Glyphosate	0.7		
Heptachlor	0.00001		
Heptachlor epoxide	0.00001		
Hexachlorobenzene	0.001		
Hexachlorocyclopentadiene	0.05		
Lindane	0.0002		
Methoxychlor	0.04		
Molinate	0.02		
Oxamyl	0.2		
Pentachlorophenol	0.001		
Picloram	0.5		
Polychlorinated biphenyls	0.0005		
Simazine	0.004		
Thiobencarb	0.07		
Toxaphene	0.003		
2,3,7,8-TCDD (Dioxin)	3×10^{-8}		
2,4,5-TP (Silvex)	0.05		

*Extra columns are for any future changes in the MCLs.

TABLE D-4 Maximum Contaminant Levels for Radioactive Elements

Constituent	MCL, pCi/L	MCL, pCi/L*	MCL, pCi/L*
Combined radium-226 and radium-228	5		
Gross alpha particle activity (including radium-226, but excluding radon and uranium)	15		
Tritium	20,000		
Strontium-90	8		
Gross beta particle activity	50		
Uranium	20		

*Extra columns are for any future changes in the MCLs.

TABLE D-5 Maximum Contaminant Levels for Microbiological Contaminants

Constituent	MCL	MCL*	MCL*
Total coliform bacteria	Presence of coliform bacteria in less than or equal to 5% of monthly samples		

*Extra columns are for any future changes in the MCL.

TABLE D-6 Maximum Contaminant Levels for General Physical Characteristics

General Physicals	MCL	Acceptable Ranges
pH	None	
Turbidity	None	
Total alkalinity	None	
Hardness	None	
Odor threshold	3 TON (threshold odor number	
Color	15 CU (color units)	

TABLE D-7 Maximum Contaminant Levels for General Minerals

General Mineral	MCL	Acceptable Range
Bicarbonate alkalinity	250 mg/L	
Carbonate alkalinity	None	
Hydroxide	None	
Total alkalinity	None	
Calcium	None	
Chloride	None	
Copper	1.0 mg/L	
Foaming agents (MBAS)	500 mg/L	
Iron	0.3 mg/L	
Magnesium	None	
Manganese	0.050 mg/L	
pH	None	
Potassium	None	
Sodium	None	
Specific conductance	900 microhms/cm	
Sulfate	250 mg/L	
Total dissolved solids	500 mg/L	
Total hardness	None	
Zinc	5.0 mg/L	

TABLE D-8 Secondary Maximum Contaminant Levels

Contaminant	Maximum Level
Chloride	250 mg/L
Color	15 CU (color units)
Copper	1 mg/L
Corrosivity	Noncorrosive
Fluoride	2 mg/L
Foaming agents	0.5 mg/L
Iron	0.3 mg/L
Manganese	0.05 mg/L
Odor	3 TON (threshold odor number)
pH	6.5–8.5 standard pH units
Sulfate	250 mg/L
Total dissolved solids	500 mg/L
Zinc	5 mg/L

APPENDIX

E

ABBREVIATIONS

°C	degrees Celsius
°F	degrees Fahrenheit
%	percent
acre-ft	acre feet
avail.	available
bhp	brake horsepower
BW	backwash
C	concentration
cm	centimeter
CT	**Concentration** of chlorine times **Time**, in minutes
D	diameter
floc	flocculation
ft	foot or feet
ft/s	feet per second
ft^2	square feet
ft^3	cubic feet
ft^3/min	cubic feet per minute
ft^3/s	cubic feet per second
g	gram
gal	gallon
GPCPD	gallons per capita per day

gpd	gallons per day
gpg	grains per gallon
gph	gallons per hour
gpm	gallons per minute
gr	grains
gr-eq	gram equivalent weight
hp	horsepower
hr	hour
HTH	high test hypochlorite
hypo	hypochlorite
in.	inch(es)
IR	inactivation ratio
kW	kilowatt
lb	pound
L	liter
LI	Langelier index
log	logarithm
m	meter
M	million, or mole (note context)
mA	milliamp
MCL	maximum contaminant level
mg	milligram
mgd	million gallons per day
mg/L	milligrams per liter
mhp	motor horsepower
mil gal	million gallons
min	minute
mL	milliliter
mol	mole
mpy	mils penetration per year
N	normality
ntu	nephelometric turbidity units

oz	ounce
pH	hydrogen ion concentration
ppm	parts per million
psi	pressure per square inch
Q	flow
r	radius
req	required
s	second
sed	sedimentation
soln.	solution
sp gr	specific gravity
temp.	temperature
TDS	total dissolved solids
UFRV	unit filter run volume
V	volume or velocity (note context)
whp	water horsepower
yr	year
ZOP	zinc orthophosphate

BIBLIOGRAPHY

American Water Works Association, *Guidance Manual for Compliance with the Filtration and Disinfection Requirements for Public Water Systems*, 1990.

California State University, Sacramento School of Engineering, *Water Treatment Plant Operation*, Volume 1, third edition, 1994.

California State University, Sacramento School of Engineering, *Water Treatment Plant Operation*, Volume 2, second edition, 1995.

Funk, Robert A., *Advanced Math Problems for Grade III/IV Water Plant Operators*, Gladstone, Oregon: Operator Certification Training, Inc., second edition, 1996.

Price, Joanne Kirkpatrick, *Applied Math for Water Plant Operators*, Lancaster, Pennsylvania: Technomic Publishing Co., 1991.

ADDITIONAL RESOURCES

Water Treatment Operator Handbook

This book covers everything water treatment operators need to know to perform their jobs and keep in compliance with changing regulations. Every phase of water treatment operator's job is addressed.

Edition: 2002, Softbound, 252 pp.
ISBN 1-58321-184-5; Catalog Number 20481

Water Distribution Operator Training Handbook, Second Edition

A comprehensive text for water distribution operator training.

Edition: 1999, Softbound, 278 pp.
ISBN 1-58321-014-8; Catalog Number 20428

Water Supply Operators I: Water Sources Textbook

Course textbook teaches the basics of the development of water sources for drinking water, groundwater, surface water, raw water characteristics, and the hydrologic cycle.

Edition: 2003, Hardback, 210 pp.
ISBN 1-58321-229-9; Catalog Number 1955.

Water Supply Operations II: Water Treatment Textbook

WSO Water Treatment provides a valuable introduction to the fundamentals of common water treatment processes and techniques.

Edition: 1995, Hardback, 523 pp.
ISBN 0-89867-789-0; Catalog Number 1956

Water Supply Operations III: Water Transmission and Distribution Textbook

Get comprehensive coverage of the basic principles behind the design, construction, operation, and maintenance of water distribution systems in WSO Water Transmission and Distribution.

Edition: 1996, Hardback, 630 pp.
ISBN 0-89867-821-8; Catalog Number 1957.

Water Supply Operations IV: Water Quality Textbook

Contains information about methods of water quality analysis and drinking water regulations.

Edition: 1995, Hardback, 252 pp.
ISBN 0-89867-804-8; Catalog Number 1958

Water Supply Operations V: Basic Science Concepts and Applications Textbook

Find the practical discussions you need of mathematics, hydraulics, chemistry, and electricity in WSO Basic Science.

Edition: 1995, Hardback, 670 pp.
ISBN 0-89867-796-3; Catalog Number 1959

Basic Chemistry for Water and Wastewater Operators

A basic chemistry primer tailored for operators of drinking water or wastewater systems.

Edition: 2002, Softbound, 178 pp.
ISBN 1-58321-148-9; Catalog Number 20494

Basic Microbiology for Drinking Water Personnel

This book provides clear, short descriptions of the waterborne microorganisms—viruses, bacteria, protozoa, and algae—which either pose a human health threat or contribute to distribution corrosion.

Edition: 2001, Softbound, 85 pp.
ISBN 1-58321-121-7; Catalog Number 20463

Operator Certification Study Guide

Prepares water and distribution operators for certification exams by providing an array of multiple choice example questions on water treatment, distribution operations, and math that operators may encounter on certification exams.

Edition: 2003, Softbound, 252 pp.
ISBN 1-58321-287-6; Catalog Number 20517

For pricing and ordering information, please visit the online bookstore at www.awwa.org/bookstore or call AWWA Customer Service at 1.800.926.7337.

www.ingramcontent.com/pod-product-compliance
Lightning Source LLC
Chambersburg PA
CBHW081124180926
51611CB00004B/48
* 9 7 8 1 5 8 3 2 1 4 5 4 1 *